# Programming the
# BeagleBone Black

D1368314

D1415927

## About the Author

**Dr. Simon Monk** (Preston, UK) has a degree in Cybernetics and Computer Science and a PhD in Software Engineering. He spent several years as an academic before returning to the industry, co-founding the mobile software company Momote Ltd. He has been an active electronics hobbyist since his early teens and is a full-time writer on hobby electronics and open source hardware. Dr. Monk is the author of numerous electronics books, specializing in open source hardware platforms, especially Arduino and Raspberry Pi. He is also co-author with Paul Scherz of *Practical Electronics for Inventors, Third Edition*.

You can follow Dr. Monk on Twitter, where he is @simonmonk2.

# Programming the BeagleBone Black

Getting Started with
JavaScript and BoneScript

**Simon Monk**

New York   Chicago   San Francisco
Athens   London   Madrid   Mexico City
Milan   New Delhi   Singapore   Sydney   Toronto

Cataloging-in-Publication Data is on file with the Library of Congress

McGraw-Hill Education books are available at special quantity discounts to use as premiums and sales promotions, or for use in corporate training programs. To contact a representative, please visit the Contact Us pages at www.mhprofessional.com.

**Programming the BeagleBone Black: Getting Started with JavaScript and BoneScript**

1 2 3 4 5 6 7 8 9 0   DOC DOC   1 0 9 8 7 6 5 4

ISBN       978-0-07-183212-0
MHID       0-07-183212-2

| | | |
|---|---|---|
| **Sponsoring Editor** | **Copy Editor** | **Composition** |
| Roger Stewart | Bart Reed | Cenveo Publisher Services |
| **Editorial Supervisor** | **Proofreader** | **Illustration** |
| Patty Mon | Carol Shields | Cenveo Publisher Services |
| **Project Manager** | **Indexer** | **Art Director, Cover** |
| Sandhya Gola, Cenveo® Publisher Services | Ted Laux | Jeff Weeks |
| **Acquisitions Coordinator** | **Production Supervisor** | |
| Amy Stonebraker | George Anderson | |

# CONTENTS AT A GLANCE

# CONTENTS

# ACKNOWLEDGMENTS

**Many thanks** to all those at McGraw-Hill Education who have done such a great job in producing this book. In particular, thanks to my editor Roger Stewart, Sandhya Goya, and to Bart Reed.

I am most grateful to Stephen Hall and Cefn Hoile for their valuable help with reviewing my efforts at JavaScript and to Robert (BobKat) Logan for his invaluable technical review.

I would also like to thank Adafruit, SparkFun, and CPC for supplying many of the modules and components used in the preparation for this book.

And last but not least, thanks once again to Linda, for her patience and generosity in giving me space to do this.

# INTRODUCTION

The BeagleBone Black combines the power of a Linux-based single-board computer with the GPIO (General Purpose Input/Output) capabilities of an Arduino.

This book shows you how to program a BeagleBone Black using its hardware library (BoneScript) and JavaScript. The book is suitable for beginner programmers and does not assume any prior programming experience.

## Downloads

The book includes many example programs, which are all open source and available from the book's website at http://beaglebonebook.com. You will need to follow the link to the code. You will also find an up-to-date list of errata for the book here.

## What Will I Need?

This is a book primarily about software. So, for most of the examples, all you really need is BeagleBone Black. However, when you are using GPIO pins, a low-cost multimeter is helpful. Male-to-male jumper leads, or even just a short length of solid core wire that can be poked into the sockets of the GPIO connector, will be needed.

The two hardware project chapters require specific parts, which are listed in those chapters.

Other parts that are recommended when learning to use the GPIO pins are a solderless breadboard and/or a BeagleBone Breadboard cape and a selection of basic components such as LEDs and switches.

Appendix A at the end of this book indicates possible suppliers for these parts.

# Using This Book

This book's main focus is on learning to program the BeagleBone Black from the viewpoint of someone who is new to programming. It therefore starts with basic concepts, gradually building on the earlier material while heading toward more advanced topics. It therefore needs to be read more or less in order.

This book is organized into the following chapters:

- **Chapter 1: Introduction and Setup**   This chapter introduces the BeagleBone Black. We look at what you can do with it as well as setting it up so that you can connect to it from your main computer over a USB or network connection.

- **Chapter 2: A Linux Computer**   The BeagleBone Black is capable of functioning as a tiny general-purpose computer, using a keyboard, mouse, and monitor. In this chapter, you will explore this aspect of the BeagleBone Black.

- **Chapter 3: JavaScript Basics**   Although you can program the BeagleBone Black in many different languages, the standard and BeagleBoard-recommended language is BoneScript, which is an extension of JavaScript. In this chapter, we get started with JavaScript programming from the point of view of the novice programmer. We will also start some simple code examples, including turning on and off the LEDs built into the BeagleBone Black.

- **Chapter 4: JavaScript Functions and Timers**   JavaScript is different from many of the languages used to program embedded electronics. It does not have a "delay" or "sleep" function. Instead, it has the concept of "timers." In this chapter, you will also start to develop an example to flash Morse code signals using the built-in LEDs.

- **Chapter 5: Arrays, Objects, and Modules**   In this chapter we get into more advanced topics and continue with the Morse code example, building it up to a general-purpose translator that will flash any message you supply as text as Morse code.

- **Chapter 6: BoneScript**   This chapter concentrates on the BoneScript library itself. You will learn how to use it to control digital and analog outputs as well as read values from analog and digital inputs.

- **Chapter 7: Hardware Interfacing**   Although this is a book primarily about programming, the basics of attaching electronics, such as LEDs, switches, servos, and so on, are described in this chapter, along with just enough electronics theory to get your project going.

- **Chapter 8: Using Capes and Modules**   This chapter is concerned with programming and using expansion capes and modules, including I2C and serial modules as well as motor controllers.

- **Chapter 9: Web Interfaces**   The web-serving capabilities of the BeagleBone Black provide a great way of making a web interface for your projects.

- **Chapters 10 and 11**   These two chapters bring together topics from throughout the book to provide end-to-end project examples. The first project is a web-controlled roving robot, and the second is an e-mail notifier that uses a 12V incandescent lamp to notify you of incoming e-mails.

## Resources

This book is supported by web pages located at http://beaglebonebook.com. You will find all the source code used in the book here, as well as other resources such as errata.

# 1

# Introduction and Setup

This chapter introduces you to the BeagleBone Black. You'll discover what it can be used for and, more importantly, what it is most suited to. You'll also learn how to connect to your BeagleBone Black over USB and immediately get started running some simple sample programs that are shipped with the BeagleBone Black.

## What Is the BeagleBone Black?

The BeagleBone Black, shown in Figure 1-1, is America's answer to the massively successful British invention, the Raspberry Pi. The BeagleBone Black is a $45 credit-card-sized single-board computer that is currently being shipped in huge numbers to hobbyists and makers keen to learn how to program the board and connect it up to external electronics.

Sitting somewhere between a Raspberry Pi and an Arduino, the BeagleBone Black is a single-board Linux computer that can be used both as a platform for embedded hardware projects and as a general-purpose Linux computer. Some of the important features of the board are:

- A 1 GHz ARM processor with 512MB of RAM.

- An operating system preinstalled to built-in flash storage (2GB). The operating system is upgradable by microSD card, or you can boot directly from the microSD card.

- A 10-second boot time. The BeagleBone Black boots up very quickly, which is important when you are using it to control electronics.

- Over 60 digital input/output pins.

1

**Figure 1-1**   *A BeagleBone Black*

- Seven analog inputs.
- An ever-expanding range of plug-in "Capes" that provide ready-made electronic expansion for LCD touch screens, battery powering your BeagleBone Black, and so on.
- An Ethernet network port.
- A micro-HDMI socket for video.
- A USB port to attach peripherals such as a keyboard and mouse.
- Low-power operation (less than 2W).

The BeagleBone Black is actually the latest in a long line of BeagleBone boards. The reason for its success is that it is much cheaper than its predecessors, being priced to compete with the Raspberry Pi. It is also a very sleek and aesthetically pleasing board.

The BeagleBoard line of single-board computers was originally designed to showcase Texas Instrument's family of System on a Chip devices, but has gathered a momentum of its own as a platform for embedded computing.

## A Tour of the BeagleBone Black

Figure 1-2 provides a quick tour of the BeagleBone Black's hardware features and connections.

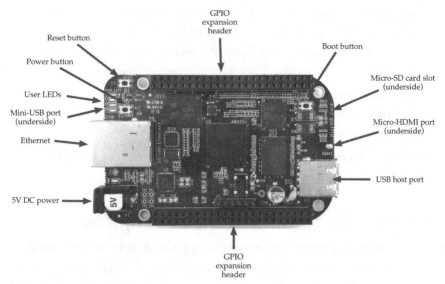

**Figure 1-2**    *The BeagleBone Black hardware features and connections*

Starting with the Ethernet socket, let's work our way around the board clockwise:

- **Ethernet port**    Allows the board to be connected to a wired network. The maximum speed is 100 Mbps.

- **User LEDs**    LEDs USR0 to USR3 can be controlled by your code, allowing you to experiment with digital outputs without having to attach external electronics.

- **Power button**    If the BeagleBone Black has been shut down from the Linux command line, pressing this button will turn the BeagleBone Black back on.

- **Reset button**    Press this button to cause a reset of the BeagleBone Black.

- **GPIO expansion header**    For attaching external electronics or expansion boards (Capes).

- **Boot button**    You only use this button when updating the flash memory of the board from a ROM image on a microSD card.

- **MicroSD card holder**    You can use an SD card to hold an image to flash the BeagleBone Black's built-in flash storage, or to boot the BeagleBone Black from it directly.

- **Micro-HDMI port**   This port allows you to connect the BeagleBone Black to HDMI using a micro-HDMI-to-HDMI lead. These leads are also used to connect smartphones to TVs and are readily available for a low cost.

- **USB host port**   You can use this port to connect a variety of USB devices to the BeagleBone Black, including a keyboard, mouse, Wi-Fi dongle, and web cam.

- **Micro-USB port**   This port is used for power and communication with your computer.

## Example BeagleBone Black Projects

If you're wondering exactly what you might like to do with your BeagleBone Black, the Internet is a great source of ideas. Many interesting projects out there use the BeagleBone Black. If you visit www.instructables.com and search for BeagleBone Black, you will find all sorts of projects, including the following:

- Environmental monitoring (measuring temperature, humidity, and so on, and displaying the information on a screen, logging it to an SD card, or providing a web interface to the data)

- Building robots of many different shapes and sizes

- Creating a home automation system

- Controlling a beer-brewing apparatus, coffee machine, and so on

## First Boot of the BeagleBone Black

That's definitely enough talk. It's now time to take the BeagleBone Black out of its elegant little box and boot it up!

### Connecting via USB

You have two ways of interacting with a BeagleBone Black. One is to connect it up to a keyboard, mouse, and monitor and use it like a regular computer. We will do this in Chapter 2. However, for most uses of the BeagleBone Black, where you use it to control electronics, it is more convenient to interact with the BeagleBone Black using a second computer (Windows, Mac, or Linux) and use a USB cable to power the BeagleBone Black as well as to communicate with the board.

The BeagleBone Black ships with a USB lead; the standard-sized USB plug end goes into a free port on your main computer and the mini-USB plug fits into the mini-USB port next to the Ethernet socket on the BeagleBone Black (see Figure 1-3).

The BeagleBone Black is ready to go, out of the box. It has an operating system (OS) installed. You will probably want to update the OS later, but for now everything is good to go, so plug the BeagleBone Black into a spare USB socket on your computer.

The BeagleBone Black will immediately start to boot up. You can tell it is doing this because the blue LEDs above the mini-USB socket will start to flash. When it has finished booting, the LED furthest from the Ethernet port will start flashing in a kind of heartbeat pattern (ba-dum… ba-dum…).

The BeagleBone Black is drawing its power from the USB connection. It will be drawing about 300 mA of the normal maximum of 500 mA that a USB port can provide. Therefore, it should be fine on just about any USB connection, as long as you do not add any extra hardware such as a Wi-Fi dongle to it.

As the BeagleBone Black boots, you may notice your operating system mounting the BeagleBone Black as a USB storage device. You can ignore this for now.

## Using the USB Network Bridge

Having booted from its out-of-the-box operating system (Ångström Linux), the BeagleBone Black runs a mini web server. It is through this web server that we are going to communicate with and program the BeagleBone Black.

**Figure 1-3**  *Connecting the BeagleBone Black using USB*

You can connect to the BeagleBone Black's web server by hooking it up to your network using an Ethernet patch cable connected to your home network hub or router. However, you do not need to do this, because those clever folks at BeagleBone have included software on the BeagleBone Black that lets its USB connection (the one we are already plugged in to) act as a network connection to your main computer.

To use this magic USB-to-network bridge, you need to install some drivers on your main computer, but only if you are using Windows or Mac OS X. If you are using Linux, you should not need to do this. It is well worth the small effort of setting up this bridge because it will allow you to use your Beagle-Bone Black with nothing more sophisticated than a USB lead.

Follow the instructions http://beagleboard.org/Getting%20Started. Scroll down to Step 2 to install the drivers for your operating system.

## Connecting to the BeagleBone Black Web Server

Once the drivers have been installed, the USB bridge will make your Beagle-Bone Black visible to your main computer on the IP address 192.168.7.2. Therefore, open a browser window on your main computer and type **http:// 192.168.7.2** into the address bar. You should see something like Figure 1-4.

At first you might think you're looking at a page served from the Internet. But check the URL bar, and you'll see that you aren't. This website is being served directly by the BeagleBone Black. What's even more impressive is that in a short while, we will be turning LEDs on and off from this browser interface.

On the left side of the web page, find the "Bonescript" section and select the link for the function digitalWrite(). On this page, you will find the example section (see Figure 1-5).

This example has a code area that already contains three lines of Bone-Script. Look closely at your BeagleBone Black's LEDs by the Ethernet port and then click the Run button. As soon as you hit the button, you should see the LED labeled USR0 light and stay lit. This is the LED that was formerly doing the little heartbeat flashing.

Edit the third line of the code by replacing **HIGH** with **LOW**. Run the code again and the LED should turn off.

You can change **USR0** to one of the other LEDs (**USR1** to **USR3**) to change the LED to turn on and off. Be sure to change USR0 on both lines to the same USR number or it will not work as expected.

**Figure 1-4**   *The BeagleBone Black web server*

Example ⊂run⊃ ⊂restore⊃

```
1  var b = require('bonescript');
2  b.pinMode('USR0', b.OUTPUT);
3  b.digitalWrite('USR0', b.HIGH);
```

Bonescript: initialized

**Figure 1-5**   *Turning on an LED from a web page*

Let's just take stock of what has happened here, because it's actually pretty amazing. The web browser on your main computer has been served a web page by the BeagleBone Black. This web page contains some JavaScript code that runs in your browser yet somehow manages to control the hardware on the BeagleBone Black.

The trick is that the example page for digitalWrite() and the example pages for the other BoneScript functions establish a connection from the browser to the BeagleBone Black, forward on the lines of code entered in the example code box to the BeagleBone Black itself, and then run them. This very powerful technique means that we can easily create a web interface to the BeagleBone Black that allows us to control hardware. Creating web interfaces like this for the BeagleBone Black is the subject of Chapter 11.

# Secure Socket Communication

To get the most out of the BeagleBone Black, you will need to use the Linux command line. SSH (Secure Shell) is a great technology that allows you to connect to the command line of a BeagleBone Black using your main computer. From the SSH command line, you can issue commands on the Beagle-Bone Black to install software, configure it, and provide other administration tasks without having to connect up a keyboard, mouse, and so on.

## The GateOne SSH Client

Back in Figure 1-4, you can see an option called GateOne SSH. This provides the web client interface for SSH shown in Figure 1-6. Current versions of the BeagleBone Black no longer offer the GateOne SSH in the navigation bar.

In theory, this will work in the browser of your main computer, and by all means try it because it is very convenient. To use it, first click the Set Date button. This sets the date and time on your BeagleBone Black to help avoid problems with certificate expiration and so on. Then click the GateOne link, and a window like the one in Figure 1-6 should be displayed.

Enter the IP address of your BeagleBone Black (192.168.7.2) and accept the default port of 22. Then enter **root** as the user. There is no password, so just hit ENTER.

The first time that you connect over SSH, you will be prompted to confirm the certificate's trustworthiness.

**Figure 1-6**  *The GateOne SSH client*

## SSH from the Mac or Linux Terminal

If you have problems with GateOne SSH, or would rather just use the command line on your Mac or Linux computer, the process is also straightforward. All you need to do is open the Terminal app on your Mac (or any shell on Linux) and enter the following command:

```
$ ssh 192.168.7.2 -l root
```

There is no password, so just hit RETURN when prompted for the password. Figure 1-7 shows a SSH shell from the Mac Terminal app.

## SSH on Windows Using PuTTY

Unlike the Mac or Linux, Windows does not come with an SSH command-line client. Fortunately, a very good open source program called PuTTY does the job. You can download PuTTY from www.putty.org.

The program is just a single file called putty.exe that you can keep on your desktop (or any convenient location). Run PuTTY by double-clicking its icon. When PuTTY starts, you'll see the configuration screen shown in Figure 1-8.

**Figure 1-7**  *SSH from the Mac (or Linux)*

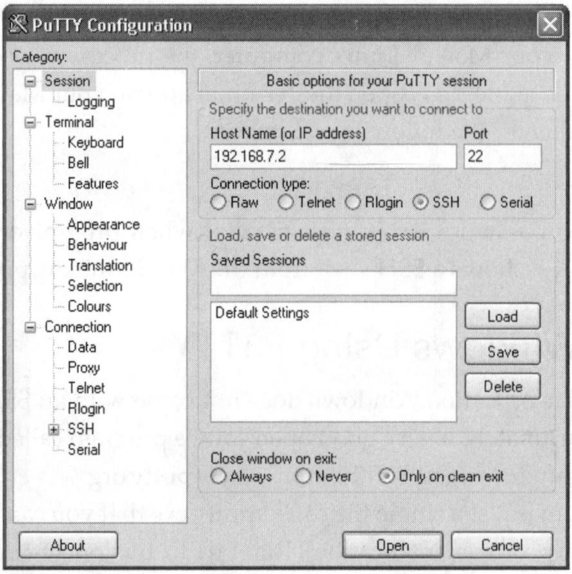

**Figure 1-8**  *Starting PuTTY*

Enter **192.168.7.2** in the field Host Name or IP Address and click the Open button. This will start a command-line SSH session on the BeagleBone Black (see Figure 1-9).

## Using the Linux Command Line

Regardless of whatever means you have used to obtain a command-line session of the BeagleBone Black, you can now type commands at the # prompt and they will be run on the BeagleBone Black. We will be returning to the command line in later chapters, so for now, let's look at a few simple commands to get us started.

The BeagleBone Black file system is organized just like the file system of a Mac or Linux computer and pretty close to that of a Windows computer. There is a root directory, which is called "/" on Linux and the Mac and called "C:/" on Windows if you have one disk partition. All the files on the computer then live in a tree structure within that root directory.

When you are using the command line, the BeagleBone Black displays the # prompt and waits for you to enter a command. So, for example, the command **ls** will list the files in the current directory.

**Figure 1-9** *SSH session using PuTTY*

When you first connect with a SSH session and log in as root, your current working directory will be the home directory for root. To confirm which directory you are currently in, type the command **pwd**:

```
root@beaglebone:~# pwd
/home/root
root@beaglebone:~#
```

So in this case, our current directory (meaning the directory we are "in") is /home/root. We can see the files and directories that are in this directory using the command **ls**:

```
root@beaglebone:~# ls
Desktop tmp
root@beaglebone:~#
```

You can change the working directory using the **cd** command:

```
root@beaglebone:~# cd Desktop
root@beaglebone:~/Desktop#
```

The Desktop folder is actually the desktop for the windowing environment that we will be exploring in Chapter 2.

## Ethernet Connection

An alternative way of connecting to your BeagleBone Black is to hook it up to your local network using the BeagleBone Black's RJ-45 Ethernet port. Use an Ethernet patch cable to connect it to a free socket on the back of your home hub/router (see Figure 1-10).

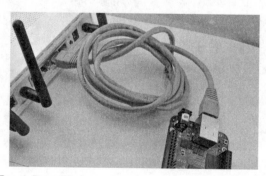

**Figure 1-10**   *A BeagleBone Black connected to a local network*

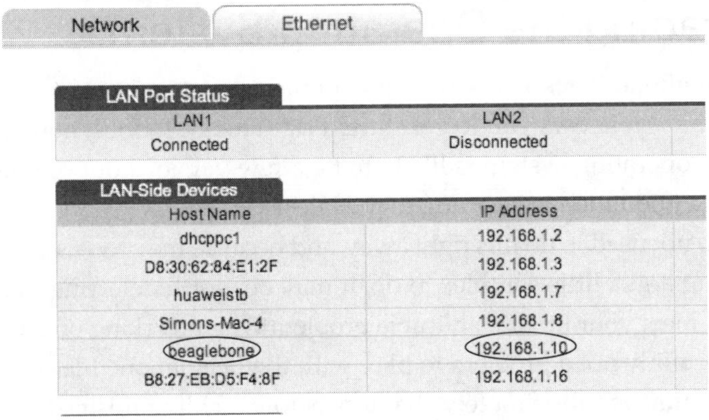

| LAN Port Status | |
| --- | --- |
| LAN1 | LAN2 |
| Connected | Disconnected |

| LAN-Side Devices | |
| --- | --- |
| Host Name | IP Address |
| dhcppc1 | 192.168.1.2 |
| D8:30:62:84:E1:2F | 192.168.1.3 |
| huaweistb | 192.168.1.7 |
| Simons-Mac-4 | 192.168.1.8 |
| beaglebone | 192.168.1.10 |
| B8:27:EB:D5:F4:8F | 192.168.1.16 |

**Figure 1-11**   *Finding the IP address of a networked BeagleBone Black*

In Chapter 2 you will also learn how to connect to a local network using a USB Wi-Fi adapter.

The IP address of the BeagleBone Black is always 192.168.7.2 when connecting using USB; however, if you are connecting with an Ethernet connection to your local network, then the network will assign the BeagleBone Black an IP address using DHCP (Dynamic Host Configuration Protocol). Depending on your network configuration, this IP address will usually start with 192.168.0 or 10.0.0. The final digit will be allocated according to the home hub's rules. For my BeagleBone Black, the IP address is 192.168.0.10.

To discover the IP address of your BeagleBone Black, you can use the administration web interface of your home hub. Log on to this (consult your home hub manual) to find the DHCP table (see Figure 1-11).

As you can see, in this case, the IP address is 192.168.1.10.

Knowing the IP address, you can now connect to the BeagleBone Black from your main computer's browser using this IP address rather than 192.168.7.2. The same is true of connecting over SSH.

Because the USB connection is now only being used to supply power, you could disconnect it and use a 5V power adapter connected to the DC barrel jack.

*CAUTION*   *It is essential that you only use a 5V power adapter with the BeagleBone Black. If you plug a higher voltage power supply into it, you will destroy your BeagleBone Black.*

# Upgrading the Operating System

The BeagleBone Black ships with its operating system preinstalled. This is great because it allows you to get started right away. It does, however, mean that your operating system is likely to be a few versions out of date. Therefore, it is a good idea to upgrade to the latest version.

You do not need to do this right away, and because the process is rather time consuming and a little complex to do, it may be worth deferring the upgrade until you meet your first inexplicable problem when working on a project.

It is certainly not a problem to play with the BeagleBone Black for a while and work through the first few chapters before tackling this task.

## Downloading the Disk Image

The first step is to download the disk image onto your main computer from http://beagleboard.org/latest-images/.

The two types of images here are the ones designed to run direct from the micro-SD card and the ones to "flash" the internal flash storage. You want the eMMC Flasher type. The image used in this book is "Angstrom Distribution (BeagleBone Black – 2GB eMMC) 2013-06-20." Because there will probably be a newer version by the time you read this chapter, you should be sure to use the latest image.

## Extracting the Disk Image

The file is downloaded as a .xz file. You may need to download a utility that can extract this type of ZIP file. For Windows and Linux, the 7-zip software (www.7-zip.org/download.html) works well. An alternative for the Mac is Unarchiver (http://wakaba.c3.cx/s/apps/unarchiver.html).

Once the ZIP file has been extracted, you will be left with an .img file.

## Installing the Disk Image on the SD Card

In an ideal world, you would just be able to copy the image file onto the microSD card; unfortunately, the image file has to be specially installed onto the microSD card to make a bootable image. The procedure for this will vary depending on the operating system of your main computer. But whatever operating system you are using, first make sure you have formatted the microSD card as FAT32 and ejected it from your main computer.

## Windows

If you are using Windows, you will need to download Win32 Disk Imager from http://sourceforge.net/projects/win32diskimager/ and then run the program (see Figure 1-12).

Click the folder icon next to the Image File field and browse to the image file. Then insert the microSD card and select it from the Device list. Finally, click Write to begin the file writing.

## Mac

If you are using a Mac, the most convenient tool for installing the image is called Pi Filler. It is actually intended for the BeagleBone Black's rival, the Raspberry Pi, but will work just as well for BeagleBone Black images.

Run the Pi Filler app and you will be prompted to browse to the location of the .img file. The app will then tell you to insert the microSD card and warn you about its impeding annihilation (Figure 1-13).

Click the Continue button and, after another warning, Pi Filler will start writing the image onto the microSD card.

## Linux

Linux also has an image writer tool that you can install using the following command:

```
sudo apt-get install usb-imagewriter
```

Run the tool, and you will see an interface similar to the Windows equivalent. Browse to the image file in the Write Image field and then insert the microSD card and select it in the To field. Finally, click the option Write to Device.

**Figure 1-12**  *Win32 Disk Imager*

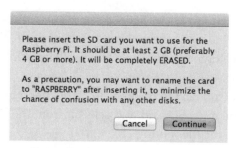

**Figure 1-13**   *A warning from Pi Filler*

## Summary

In this chapter you learned how to set up your BeagleBone Black and connect to it using the USB cable provided, as well as how to connect to the device using an Ethernet cable. All of this is done without needing to hook up a keyboard, mouse, and monitor to the BeagleBone Black.

In the next chapter we will look at how you can use the BeagleBone Black as a stand-alone computer with those peripherals attached.

# 2

# A Linux Computer

In addition to acting as an embedded controller, the BeagleBone Black can also be used as a regular Linux computer. In this chapter, we explore this aspect of the board's personality.

## Attaching the Keyboard, Mouse, and Screen

The BeagleBone Black has only one USB port (the type to which you attach a keyboard and mouse). Therefore, you will need to buy a USB hub to allow both items to be plugged in at the same time, as well as any other peripherals you may need. A more compact alternative is to use a wireless keyboard and mouse combo, which comes as a kit containing a single USB dongle that then connects wirelessly to a keyboard and mouse. Note that the Bluetooth variety of wireless keyboard and mouse are a lot more trouble than the non-Bluetooth variety.

To attach the BeagleBone Black to a monitor or TV with an HDMI input, you will need a mini-HDMI-to-HDMI adapter, shown in Figure 2-1, plus a regular HDMI lead, or you can buy a lead that has a mini-HDMI plug on one end and a full-size HDMI plug on the other end. The latter solution is probably better because the adapter is quite bulky and sticks out below the rest of the board. Thus, if the whole thing were pressed down accidentally, it could lever off the mini-HDMI socket from the circuit board.

With everything plugged in, power up the BeagleBone Black, and after a while you should see something like Figure 2-2. Note that you need to plug

**Figure 2-1** *BeagleBone Black with an HDMI adapter*

in the HDMI cable before booting; if you plug in the HDMI cable after booting, the BeagleBone Black will not detect that a monitor is attached and therefore will not activate the video hardware.

You should also now connect your BeagleBone Black to the Internet via your network. You may be able to do this through your USB connection to your main computer, if your main computer has Internet sharing turned on. However, the most foolproof way is to use an Ethernet patch cable and connect it directly to your home hub. Later on in this chapter, we will set up the BeagleBone Black to use a USB Wi-Fi adapter.

**Figure 2-2** *First boot screen*

# Setting the Date

Unlike most full-size computers, the BeagleBone Black does not have a real-time clock. Nor does it automatically set its system time and date from the Internet like the Raspberry Pi does. In addition to displaying the wrong time on the desktop, an incorrect date on a computer generally causes problems in a number of areas. Web browsing, installing software, and connecting to other computers are all likely to be problematic without the correct date. This has to do with security certificates that have valid date ranges. Therefore, if the computer does not have the right date, the certificates we need to use may well be invalid.

To set the time, open the Terminal application, shown in Figure 2-3, from the System Tools submenu of the Applications menu.

Type the following command into the Terminal window and press RETURN:

```
# ntpdate -b -s -u pool.ntp.org
```

After a short delay, during which the screen may flicker off, you will notice that the date and time shown in the top right of the screen are now correct. You may have to wiggle the mouse and press some keys to get the computer to wake.

To avoid having to run this command every time we reboot, we can arrange for the preceding command to be run as the BeagleBone Black boots up.

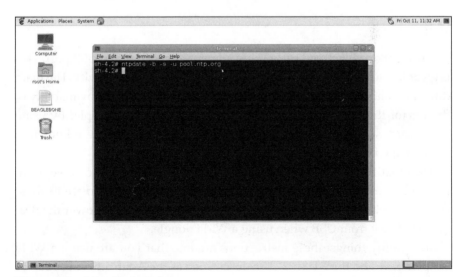

**Figure 2-3**   *A Terminal window*

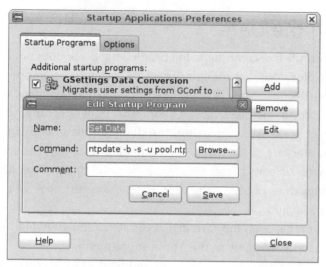

**Figure 2-4**  *Setting the date at startup*

The simplest way to do this is to use the Startup Application tool, which can be found on the System Menu under Preferences.

Click the Add button and in the Name field enter **Set Date**. In the Command field, paste the full text of the **ntpdate** command and then click the Save button (see Figure 2-4).

# Wi-Fi

The BeagleBone Black can use a Wi-Fi adapter attached to the USB port. This is great for projects, where you need wireless Internet connectivity. The main difficulty with setting up Wi-Fi on the BeagleBone Black is the availability of drivers for the Wi-Fi hardware. This process can be made simpler by using Wi-Fi hardware that contains a chip for which Linux drivers are known to work on the BeagleBone Black.

The other thing you need to be aware of is that Wi-Fi is quite power hungry, and unless you use a 5V 1.5A (or more) power supply, you are likely to encounter reliability problems. Therefore, do not try and just power the BeagleBone Black from USB when using a Wi-Fi dongle.

To simplify things, these instructions assume that you are using a Wi-Fi adapter based on the RTl8192CU chipset, such as the Adafruit Wi-Fi adapter listed in Appendix A.

The "conman" connection manager will use a wired network connection or USB network connection in preference to a Wi-Fi connection. This means that you cannot do the entire Wi-Fi setup using SSH over Ethernet or USB. You will need to have a keyboard, mouse, and monitor attached to the BeagleBone Black at least for some of the procedure.

## Step 1: Upgrade

It is a good idea to upgrade to the latest disk image for the BeagleBone Black (refer to Chapter 1). Before you do, remember to copy any files you need off the BeagleBone Black so as not to lose them when you wipe the SD card.

You may choose to try the procedure without updating and then update if it fails.

## Step 2: Install the Driver

To perform this step, you need an Internet connection, so plug in the Ethernet patch cable. You can carry out this step entirely over SSH or directly in a Terminal window on the BeagleBone Black. Also, make sure the date is set correctly (see the previous section in this chapter).

Plug in the USB Wi-Fi dongle. If you are going to use a keyboard and mouse, you will need a USB hub.

Regardless of whether you have an SSH session or a direct Terminal window on the BeagleBone Black, type the following commands:

```
# opkg update
# opkg install linux-firmware-rtl8192cu
# reboot
```

You can now check that the USB dongle drivers are working using this command:

```
# lsusb
Bus 001 Device 002: ID 0bda:8176 Realtek Semiconductor Corp.
RTL8188CUS 802.11n WLAN Adapter
Bus 001 Device 001: ID 1d6b:0002 Linux Foundation 2.0 root hub
Bus 002 Device 001: ID 1d6b:0002 Linux Foundation 2.0 root hub
#
```

The important thing to look for here is "RTL8188CUS 802.11n WLAN Adapter." This tells you that the driver is running okay. Depending on your hardware, the message may not be exactly the same. The important part is that it mentions 802.11n or WLAN or Wi-Fi.

## Step 3: Configure

You can stick with the SSH connection for this step, if you like. Here, you need to create a new file using the following command:

```
# nano /var/lib/connman/wifi.config
```

Now enter the following text into the editor window and then save it using CTRL-X followed by Y:

```
[service_home]
Type = wifi
Name = my_wifi_network_name
Security = wpa
Passphrase = my_wifi_password
```

Replace my_wifi_network_name with the name (SSID) of the wireless network you want to connect to, and replace my_wifi_password with your Wi-Fi password.

## Step 4: Test

At this point, you cannot have the Ethernet cable attached. Therefore, unplug your BeagleBone Black and then attach the keyboard, mouse, and screen. Keep the Wi-Fi adapter plugged into the hub and let your system boot up again.

Open a terminal session and type the following command:

```
# ifconfig wlan0
wlan0     Link encap:Ethernet  HWaddr 00:E0:4C:12:A2:9C
          inet addr:192.168.1.6  Bcast:192.168.255.255
          Mask:255.255.0.0
          inet6 addr: fe80::2e0:4cff:fe12:a29c/64 Scope:Link
          UP BROADCAST RUNNING MULTICAST  MTU:1500  Metric:1
          RX packets:2152 errors:0 dropped:0 overruns:0 frame:0
          TX packets:898 errors:0 dropped:0 overruns:0 carrier:0
          collisions:0 txqueuelen:1000
          RX bytes:885461 (864.7 KiB)  TX bytes:98986 (96.6 KiB)
```

You can see that in this case, the BeagleBone Black has been assigned an IP address of 192.168.1.6 on the wireless network.

At the time of writing, there are still some reliability issues with getting Wi-Fi working. For instance, my Wi-Fi adaptor will only work when attached through a hub. If the Wi-Fi adaptor is plugged directly into the USB port, even though "lsusb" recognizes it, it will not join my network. Put a very cheap and old USB hub (not even a USB2 hub) in there, and it works just fine!

Hopefully by the time you read this, the problem will be resolved. So if you have trouble when you follow these instructions, a quick Internet search may put you straight.

## Bundled Software

If you booted your BeagleBone Black with a monitor attached, you will see the Applications menu across the top of the screen that provides access to a number of built-in applications. Under its Accessories submenu you can find a text editor called "gedit," and under the Graphics submenu you can find the GNU Image Manipulation Program (better knows as GIMP).

The Internet menu has links to various browsers, as shown in Figure 2-5, and also a program called X11VNC Server. If you run this option, it appears that nothing has happened, but in the background a VNC (Virtual Network Connection) is running that is sharing your computer screen over the network. To view it, you need a VNC client such as VNC Viewer (http://www .realvnc.com/download/viewer/). When you're asked to connect, enter the IP address of your BeagleBone Black, followed by **:0**.

Under the Office submenu you will find Abiword, which, as the name suggests, is really a word processor.

Another useful program is the file browser, shown in Figure 2-6. You can find this program in the System Tools submenu.

**Figure 2-5**  *Internet programs pre-installed on the BeagleBone Black*

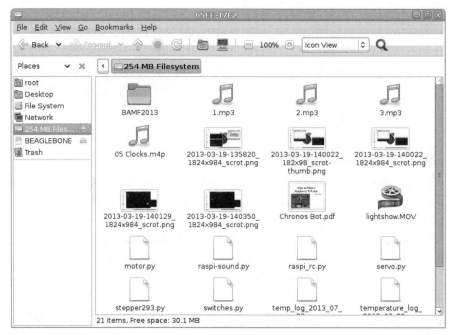

**Figure 2-6** *The File Browser*

The file browser allows you to move files around as well as organize and delete them. If you plug a USB flash drive into a USB socket, it will appear on the left side of the window. You can use this to get files on and off your BeagleBone Black.

# Installing More Software

New applications for the BeagleBone Black are installed over the Internet using the "opkg" package manager. This is a command-line tool you run from a Terminal window or over SSH. The package manager relies on an up-to-date list of available packages. So you should run the following command to update the lists before downloading new software:

```
# opkg update
```

Let's say we wanted to download the Gnumeric spreadsheet software to accompany the Abiword word processor we already have. To do this, issue the following command:

```
# opkg install gnumeric
```

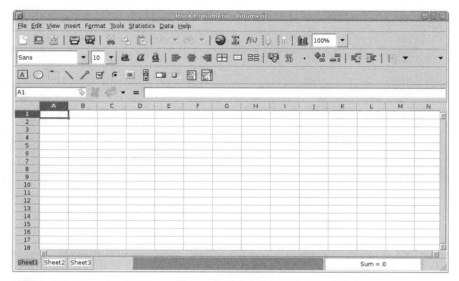

**Figure 2-7**   *Running the Gnumeric spreadsheet on BeagleBone Black*

After a while, the software will have been installed and you can find it in the Application menu. Gnumeric is shown in Figure 2-7.

Many other programs can be downloaded and run on the BeagleBone Black. However, remember that your entire file system is only 2GB in size, so you will quickly run out of room.

## Summary

The BeagleBone Black makes quite a useable Linux computer for simple tasks such as word processing or web browsing. However, the real power of the device lies in its ability to control hardware and become part of a system of your design.

In the next chapter, you will get to grips with programming the Beagle-Bone Black.

# 3

# JavaScript Basics

The BeagleBone Black can be programmed in many languages; however, the official and easiest language to start with is BoneScript. In this chapter, we look at the programming environment and also the basics of the JavaScript language used by BoneScript.

## Programming

One of the great things about the Maker Revolution is that people from all sorts of backgrounds, with different levels of technical expertise, are not afraid to take a building block like the BeagleBone and use it to create something interesting.

Many people who want to use a BeagleBone Black come from an electronics background, or they are a "maker" who has an invention in mind and wants to build it, and is prepared to learn enough technology to accomplish this. This chapter is written for those people. Therefore, if you have programmed before, you may find you know a lot of this material already.

## Operating System

A BeagleBone Black is essentially a computer. Like all computers, it will not do anything unless it is given instructions. These instructions are its software, or programs that "run" on the computer and have it do things.

When you first plug in your BeagleBone Black, LEDs will flash, and if you have a monitor, keyboard, and mouse attached, you will see a Windows-style desktop. This is because the BeagleBone Black comes with software pre-installed onto its built-in flash storage. This built-in software is called the "operating system," and for a BeagleBone Black that is fresh out of the box, this operating system will be a variety of Linux called Ångström.

The operating system boots up when power is first applied to the board; once it has started up, the operating system performs a number of roles:

- It allows the computer hardware to appear to run lots of programs at the same time. This is called *scheduling*. It allows each of the running programs (or processes) to have a share of the CPU's time (CPU is short for central processing unit). Thus, one program will get a couple of milliseconds, then another gets the CPU's time, and so on.

- It provides a user interface for interacting with the computer. This is the windowing system you see when you have a monitor attached.

- It provides an interface to peripheral hardware such as the monitor, keyboard, and mouse.

- It provides a means for users to install and run other programs—either ones they have downloaded or ones that they have written themselves.

Figure 3-1 illustrates the software of the BeagleBone Black as a series of concentric circles around the central core of the actual hardware of the board.

The layer above the actual hardware is the device drivers layer. Here, you will find the programs responsible for interfacing with hardware components such as USB devices, video output, Wi-Fi adapters, the GPIO hardware, and more. These are available to the layers above, such as the shell (that this, the Linux command line) and the windowing system. Also using the shell layer are utility programs and daemons. Daemons are monitoring programs that run all the time in the background, responding to events that happen. For example, there is a daemon watching for USB devices to be plugged in and a daemon for spotting changes in the network connection.

Above this layer is the windowing system, which allows you to interact with the computer using a mouse rather than simply using a keyboard and the command line.

Finally, the outermost layer is where the applications and programs you write will live.

## What Is a Program?

A program is a set of instructions written in a programming language. Once the program has been written, the computer (in this case, the BeagleBone

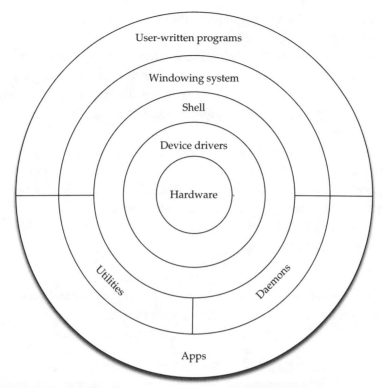

**Figure 3-1** *Linux as Layers*

Black) can be told to run the program—or, more accurately, the operating system of the computer can be told to run this program.

*NOTE* *One of the key differences between the BeagleBone Black and the Arduino board is that the Arduino board has no operating system. You just install one program on it and that is the program that runs when the board is powered up.*

Programs are written as text into a file. There are various names for this file of program instructions, including "source code," "source," "code," "program code," and "script."

## Programming Environment

Because the code for a program is just text, you can actually write it in any program designed for editing text. Therefore, you will be using a program that someone else has written to write the code for your program. This is all a bit "which came first, the chicken or the egg?"

You can write your code in a simple text editor such as nano, which we will discuss soon. However, it is much easier to write your code using a special-purpose "program editor" or IDE (integrated development environment). This keeps everything you need while writing and testing the program within one environment.

In the next section, we look at the Cloud9 IDE, which allows you to program the BeagleBone Black from another computer without having anything more than a USB lead connected to the BeagleBone Black. But for now, let's try writing and running a little program using nano.

Open a Terminal window or SSH session on the BeagleBone Black and type the following command:

```
# nano hello.js
```

Type the text shown in Figure 3-2 into the editor window and press CTRL-X and then Y followed by the ENTER key to save the file.

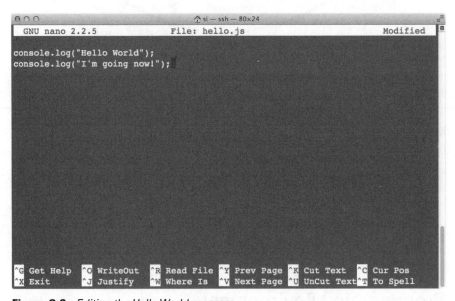

**Figure 3-2**  *Editing the Hello World program*

You can now run the program as shown here:

```
# node hello.js
Hello World
I'm going now!
root@beaglebone:~#
```

The program has followed our two lines of instructions to display two lines of text. The program has now run out of things to do and therefore gives us control back at the # prompt.

To run our program, we actually used another program called "node" and told it the name of our program file, which it then ran.

# Cloud9 Web IDE

Although the BeagleBone Black functions pretty well as a general-purpose Linux computer, for a lot of applications the BeagleBone Black will be used to control some electronics using its GPIO ports. For such applications, the keyboard, mouse, and monitor needed to program the board are an unnecessary encumbrance. Therefore, it is useful that we can program the board without having to attach these peripherals.

As the name suggests, the Cloud9 Web IDE, shown in Figure 3-3, is a web- or network-based IDE. Whereas most IDEs are stand-alone programs, Cloud9 runs in a browser. What's more, it can be a browser on any computer on the same network as the BeagleBone Black. This means you can program the BeagleBone Black from the comfort of your usual PC, Mac, or Linux computer using just a browser.

As you saw in Chapter 1, when you connect a BeagleBone Black to your computer using USB, it appears to have a network connection. Therefore, you can aim a web browser at the URL http://192.168.7.2 and the BeagleBone Black will serve a set of web pages back to you.

If you specify a port of 3000 (by adding **:3000** to the end of the preceding URL), you will see the Cloud9 IDE running on the BeagleBone Black, all ready for you to start writing some programs. This is the URL used in Figure 3-3.

Cloud9 will work with most recent browser versions. For the purposes of this book, we will be using Google Chrome.

On the left side of the Cloud9 screen is a list of project files. These are grouped in a folder called **cloud9**. To edit one of these files, double-click the filename in the Project Files area and it will open up for editing in the editor area.

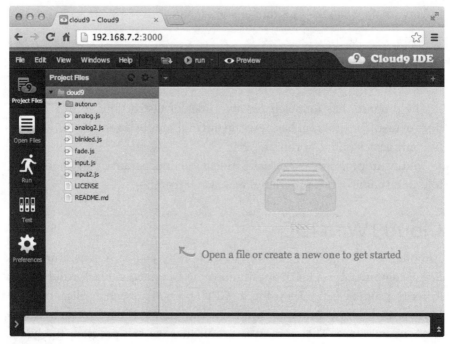

**Figure 3-3**  *The Cloud9 web IDE*

At the top of the Cloud9 window you will find a series of menus. Take some time to look at these different menu options to get an idea of the features available in Cloud9. Along the far left of the window is a series of icons that show alternatives to the list of project files. For example, you can see a list of preferences for the IDE.

Let's now revisit our "Hello World" example, but this time using Cloud9 instead of nano.

Because we will be creating quite a few program files in Cloud9 as we work through this book, let's start by creating a folder in which to keep them all. To do this, right-click over the cloud9 folder icon in the Project Files area and select the option New Folder. Name the new folder **Prog BBB**. Next, right-click over the newly created folder, select the option New File, and name the file **hello.js**. When you do this, it may appear that nothing has happened. You will not actually see the new file until you open the Prog BBB folder by clicking it.

You can rename files and folders at any time, simply by clicking their filenames in Project Files.

Double-click the new file in the Program Files area so that a blank editor for the file appears on the right side of the window; then type the following text into the newly created file:

```
console.log("Hello World");
console.log("I'm going now!");
```

The Cloud9 window should now look something like Figure 3-4.

You will notice that an asterisk (*) appears next to hello.js. This indicates that the file has not yet been saved. You should save the file before running it; otherwise, the previously saved version of the file will be run. The quickest way to save the file is to press CTRL-S (or CMD-S if you are a Mac user).

You do not really need to know just where Cloud9 is saving your files. They will still be there next time you connect to Cloud9. However, for the interested, they are saved in /var/lib/cloud9.

To run the program, click the little green triangular button marked Run. When you do, the window will automatically adjust to show the Output area, where you can see the results of the program being run (see Figure 3-5).

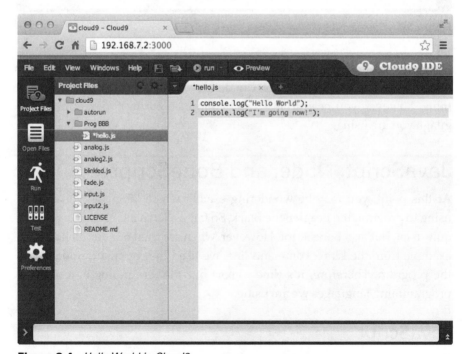

**Figure 3-4**   *Hello World in Cloud9*

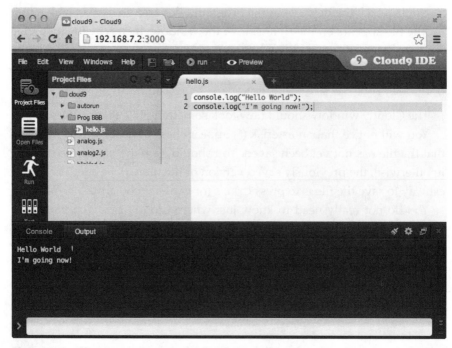

**Figure 3-5** *Showing the output of Hello World*

You will meet many other interesting features of Cloud9 as you work your way through the book. Before you can do much more in the way of programming, however, you need to learn some more about the programming language you are using.

# JavaScript, Node, and BoneScript

At this point, you may be wondering exactly which language you will be using to program the BeagleBone Black. So far, we have mentioned JavaScript quite a lot, but also BoneScript. However, when we tried out our Hello World example from the Linux command line, we used the command **node** to run the program. Therefore, it's time to sort out the terminology around the programming languages we are using.

## JavaScript

Primarily we are programming in the JavaScript language. This programming language was made popular by its use in web browsers. Traditionally, you might have embedded some JavaScript in a web page to, say, validate a field or disable a Submit button after it has been clicked. In recent times, though, the role of JavaScript has expanded to the extent that it handles many tasks on a web page, often including communication back to the server, without the reloading of the entire web page.

As essentially a language designed to run in a browser, JavaScript may seem like a strange choice for programming a BeagleBone Black. However, JavaScript has the great advantage that a whole host of website programmers are familiar with it, and it is a pretty easy language to get started with.

Note that, unsurprisingly, people often confuse JavaScript with the programming language Java. They are totally separate languages and are actually completely different in many important ways.

## Node

Clearly, the Linux command line is not a browser, yet earlier on we ran some JavaScript from it. This is where node (more properly called Node.js) comes in. Node is a program that allows you to run JavaScript programs without a browser. It is built on top of Google's V8 JavaScript technology. This is how we were able to run our Hello World JavaScript code from a command line, without any browser being present.

As well as providing a way to run JavaScript programs without a browser, node also has built-in features for networking that make it very simple to create web server projects. There is no better illustration of this than the Cloud9 IDE, which is itself written in node.

## BoneScript

This brings us around to BoneScript. BoneScript is actually just a bunch of code that allows us to easily interface to the GPIO pins of the BeagleBone Black using simple JavaScript commands. These commands allow us to turn output pins on and off and read digital input pins. It also allows for the reading of analog inputs so that we can measure voltage (say, from a sensor).

# Experimenting

The best way to get a feel for a programming language is to experiment. Node provides a facility to enter JavaScript commands on a special

**Figure 3-6**  *Using node interactively*

command line and see the results immediately. You can use this feature by connecting to a BeagleBone Black using SSH (see Chapter 1) and then typing **node** (see Figure 3-6).

Anything you type here will be evaluated (run) when you press ENTER. So, for example, typing **2 + 2** will be evaluated as a JavaScript expression and the result **4** will be displayed on the line below.

While you are learning the basics of JavaScript, you could also just download node for your main computer from http://nodejs.org. However, sticking with the BeagleBone Black, we will use Cloud9 and create a file into which we can type a command or two and then run them. This may as well live in the Prog BBB folder we created earlier. Create a new file here called **play.js**.

# Numbers

Using numbers and arithmetic in JavaScript is very much like in the world outside of programming. Therefore, if we want to add 2 and 2 together, we can just type **2 + 2**. However, to be able to see the answer, we need to write the result to the Console. The Console in Cloud9 is the area at the bottom of the screen that pops up whenever there is anything to be displayed. We write to the Console using the command **console.log**. The command needs to know what it is displaying, which is contained between parentheses. Finally, the end of the line of JavaScript is marked with a semicolon.

Therefore, type **console.log(2 + 2);** into your newly created play.js file, save it, and then run it. Figure 3-7 shows the result.

Here, we have asked JavaScript to add 2 and 2, and it has told us the answer is 4 in the Console. Numbers like this (whole numbers) are referred to as *integers* by programmers.

Suppose you need to use numbers that have a decimal point in them. These types of numbers are called *floats* or *floating-point numbers*. We could easily use a float in our simple example. Try adding the line **console.log(2.0 + 2.0);** to play.js and then run it again. This time the console will display 4 twice.

Now change the second float to **2.5** and then save the file and run it. You will see the result 4.5.

Javascript is unusual amongst programming languages as it does not distinguish between integers and floats. This means that 4 and 4.0 are the same thing in Javascript.

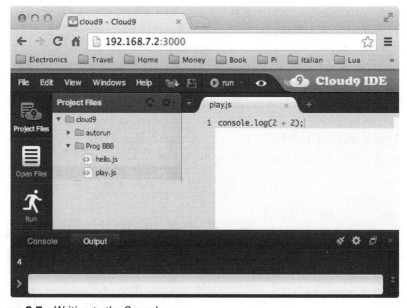

**Figure 3-7** *Writing to the Console*

# Variables

In programming, the word *variable* refers to giving a name to a value. Try changing play.js to the following and run it:

```
var x;
x = 2;
console.log(x);
```

The same result of 2 will be displayed in the Console.

The word **var** tells JavaScript that we are defining a variable. This is then followed by the name we are to give this variable (in this case, **x**). The next line gives the variable a value of 2. The final line writes **x** to the Console. Because **x** has the value 2, it is 2 that will appear in the Console.

To save yourself a bit of typing, you can combine the two lines that define the variable and give it its initial value into a single line, like this:

```
var x = 2;
console.log(x);
```

Note that JavaScript is not much interested in how things are formatted and how many spaces there are between words—just as long as there is at least one space between words. It does not even need a space for special characters such as the equals sign (=). You could, for example, write the lines of this example as follows:

```
var x=2;console.log(x);
```

Although JavaScript does not care how our code looks, we do. Nicely formatted code is much easier to understand. Therefore, you should space things out a little.

Having created a variable, you can always assign a new value to it using the = operator. Try out the following example:

```
var x = 2;
x = x + 1;
console.log(x);
```

This displays the answer 3, because we have set the value of **x** to be whatever **x** was before (2), plus 1. Note that after the first declaration of **x** we do not need to put **var** in front of it again.

In these examples, we have used very simple variable names (in fact, single-letter names). This is fine for small programs or where the purpose of the variable is obvious from the context. However, programs are not always that simple, and using longer more meaningful names makes your code much easier for others (as well as yourself, after a few days of doing something else) to understand.

By convention, variable names in JavaScript start with a lowercase letter. The first letter can then be followed by any letter (upper- or lowercase) or digit or an underscore character (_). When you have a variable name that is made up of more than one word, you cannot uses spaces. The convention is to start each new word after the first word with an uppercase letter. This way of separating words in variable names is called *bumpy case* or sometimes *camel case*, for obvious reasons. For example, the following are examples of good variable names:

- counter
- numberOfDice
- x1

And here are some examples of legal but nonstandard names for variables:

- COUNTER
- Counter
- number_of_dice
- x_1

Finally, here are some examples of illegal variable names:

- 1x
- number_of_dice!

## Strings

In addition to manipulating numbers, a programming language must be able to manipulate text. In programming-speak, pieces of text are called *strings* (short for *character strings*). In JavaScript, strings are enclosed in either single or double quotes.

We can give a variable a string value as follows:

```
var s = "Programming";
console.log(s);
```

When concocting messages to be displayed, it is quite common to want to join one string to another, or *concatenate* the two strings (more programming terminology). We can join together strings using the + operator. Here's an example:

```
var s1 = "Programming";
var s2 = "BeagleBone Black";
var s = s1 + s2;
console.log(s);
```

The output in the Console for this will be as follows:

```
ProgrammingBeagleBone Black
```

We are missing a space between s1 and s2. One way to fix this is to change the example, as shown:

```
var s1 = "Programming";
var s2 = "BeagleBone Black";
var s = s1 + " " + s2;
console.log(s);
```

The output in the Console will now be this:

```
Programming BeagleBone Black
```

## Dice Example

The instructions we have been using in our example code so far are frankly dull. Let's introduce a bit of randomness into the proceedings. We can ask JavaScript to generate random numbers for us and use this feature to display a random number between 1 and 6 as if we were rolling a die.

Type the following into play.js and then run it:

```
var x = Math.random();
console.log(x);
```

The result displayed in the Console will be a float greater than or equal to 0 and less than 1. Run the program a few more times, and you will see that you get a different number each time.

Converting a floating point number with many decimal places into an integer between 1 and 6 requires a bit of math. Let's start by multiplying the number by 6. Run the following code a few times:

```
var x = Math.random() * 6;
console.log(x);
```

*NOTE*   *As with most programming languages, the * character is used to denote multiplication.*

So, now we are getting numbers between 0.0 and 5.9999999999. We can cut off the places after the decimal point using the command **Math.floor**. The number you wish to extract the "whole number" from is enclosed in parentheses. Therefore, in this case we can assign a new value to the variable of the old value of **x** floored:

```
var x = Math.random() * 6;
x = Math.floor(x);
console.log(x);
```

Run it a few times and you will notice that we are getting numbers between 0 and 5 rather than 1 and 6. Therefore, we have the right span of numbers, just not quite in the right range. This is easily fixed by just adding 1 to the result of the **Math.floor** operation:

```
var x = Math.random() * 6;
x = Math.floor(x) + 1;
console.log(x);
```

If we were playing *Monopoly*, or for that matter *Snakes and Ladders*, it might be useful to be able to throw two dice at a time. Therefore, let's double up the code and call our other die **y**:

```
var x = Math.random() * 6;
x = Math.floor(x) + 1;
console.log(x);

var y = Math.random() * 6;
y = Math.floor(y) + 1;
console.log(y);
```

Now, every time we run the program, we will see two numbers in the console rather than just one. However, if we want to know the total of the two scores, we have to add it up ourselves. Therefore, let's modify the program again so that it does the addition for us:

```
var x = Math.random() * 6;
x = Math.floor(x) + 1;
console.log(x);

var y = Math.random() * 6;
```

```
y = Math.floor(y) + 1;
console.log(y);

var total = x + y;
console.log("Total: " + total);
```

## Ifs

Our program is getting a little more interesting now. To continue, let's say we want to automatically check for a "double" being thrown. Thus far, all our programs have run each line of code, in turn, just once and then quit when there was nothing left to do. A fundamental part of programming is the concept of conditional branching. In other words, sometimes you only want to run a certain chunk of code if some condition is true. Therefore, we would only want to print a message that a double has been thrown when the rolls of the two dice are the same number. The key word here is *if*. Thus, the **if** command in JavaScript can be used to make the running of a particular line (or lines) of code conditional.

Edit the contents of play.js to match the following lines:

```
var x = Math.random() * 6;
x = Math.floor(x) + 1;
console.log(x);

var y = Math.random() * 6;
y = Math.floor(y) + 1;
console.log(y);

var total = x + y;
console.log("Total: " + total);

if (x == y) {
    console.log("DOUBLE");
}
```

Run the program a few times and, eventually, you will throw a double and be rewarded with the message "DOUBLE." Let's now examine the syntax of these new lines.

Immediately after **if** is an expression enclosed in parentheses. This is called the *condition* of the **if** statement. At first sight, this condition (**x == y**) looks a little like we are assigning a value to **x**. However, this is not the case, because we are using double equals (==) rather than just a single (=). The double equals is a comparison operator. That is, it has a value on either side of it, and it compares those two values. If they are equal, the **if** statement will run all the commands inside the curly braces (that is, { and }). By convention, the first curly brace is placed on the same line as the **if**.

# Looping

Aside from **if**, the other mainstay of programming is the ability to repeat sections of the code a number of times. Let's start with the simple example of making JavaScript count to 10. Enter the following code into play.js and run it:

```
for (var i = 0; i < 10; i++) {
    console.log(i);
}
```

You should see something like what's shown in Figure 3-8.

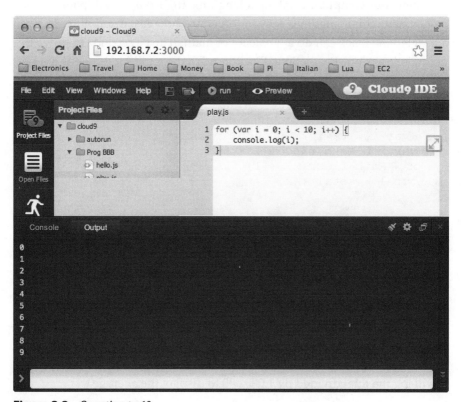

**Figure 3-8**  *Counting to 10*

The **for** command allows us to run all the code between the curly braces a number of times. The stuff inside parentheses after the **for** command determines just how many times. This code is split into three parts, separated by semicolons. The first part (**var i = 0;**) defines a variable called **i** and gives it the initial value of 1. This just happens once, before the code in the loop starts running.

The next section (**i < 10;**) is the condition for staying inside the loop. This is just like the condition we used in **if**, except in this case the condition is that **i** is less than 10.

The final section (**i++**) uses the special command **++** to add 1 to the variable that proceeds it. You could, if you prefer, write **i = i + 1** instead of **i++**. It just saves a bit of typing.

The things we want to repeat are contained within the curly braces belonging to the **for** command. In this case, we will run the single line of code to print the value of the variable **i** in the Console output window.

When your computer is taking a long time to display a web page or open a document, it is sometimes easy to forget just how fast a computer is. Even a BeagleBone Black can do a whole lot of math in just a second. To prove this point, let's ask our BeagleBone Black to roll a die a million times and report back the average value rolled. Try out the following code:

```
var total = 0;

for (var i = 0; i < 1000000; i++) {
    var x = Math.random() * 6;
    x = Math.floor(x) + 1;
    total = total + x;
}

console.log(total / 1000000);
```

You should see an answer close to 3.5. Equally astonishing is that the BeagleBone Black will probably take less than a second to come back with the answer and display it.

The program uses a new variable called **total**, which is defined before we get to the **for** loop. Inside the **for** loop, we roll a single die, just like we did earlier, but instead of writing the value to the Console, we add it to the **total** variable.

Once the loop has finished its work (that is, when **i** is equal to 1,000,000), it will log the total divided by 1,000,000 to display the average.

# Turning on USR LEDs

The BeagleBone Black has a row of four LEDs near the Ethernet port labeled "User LEDs." Next to each LED is a number, 0 to 3. The out-of-the-box configuration of the BeagleBone Black will have these LEDs flashing in various ways; in particular, LED 0 will flash in a heartbeat-type pattern.

These LEDs can be controlled using JavaScript. What makes it possible to control these is BoneScript. BoneScript is a JavaScript library. A *library*, in this case, means a bundle of code someone else has written that will allow us to interact with the BeagleBone Black hardware, including these LEDs, using JavaScript commands.

Try out the following example to stop that annoying blinking of LED 0:

```
var b = require('bonescript');

b.pinMode("USR0", b.OUTPUT);
b.digitalWrite("USR0", 0);
```

LED 0 will now be off. Change the **0** on the last line to **1** and run the program again. The LED should now come on and stay on.

To make use of the "bonescript" library, we need the top line of code that "imports" the library so that we can use it in our JavaScript programs. The result of importing it is assigned to a variable. This is interesting because we can be fairly sure that whatever the result of importing the library is, it is not just a simple string or number, as we have used with variables so far. Actually, it's what is called an "object," and you will learn more about objects later in the book.

The other two lines of the program both use this object. The second line of the program uses a command associated with **b** called **pinMode**. The period between **b** and **pinMode** shows that the **pinMode** command belongs to **b** and hence to the bonescript library. The command sets the mode of the LED to be an output. In the case of an LED, the connection could not really be anything but an output, but the command is general purpose and can also be used to set any of the pins on the GPIO sockets to be inputs or outputs.

The last line of the program writes a value to that output of either 0 (off) or 1 (on).

## Summary

In this chapter we started programming in JavaScript using the Cloud9 IDE and node. We also had the briefest of encounters with BoneScript, turning on and off one of the built-in LEDs on the BeagleBone Black.

In the next chapter we will build on this foundation, and you will learn a lot more about JavaScript.

# 4

# JavaScript Functions and Timers

**In this** chapter, you'll discover a lot more about JavaScript. We also look a bit deeper into Cloud9, and you'll learn how to use it for debugging JavaScript.

This chapter discusses the more advanced features of the JavaScript language such as timers, objects, and modules. Later on in the chapter, the program examples are long enough for you not to want to type them in from scratch. You can find the code for all the programs (of any length) in the book at the book's web page (www.beaglebonebook.com) under the Code section. Here, you will find each program in its own file, identified by the chapter number and then the program listing number.

## More on Ifs

In the previous chapter, we skated over the surface of some of the concepts in the interest of getting a quick start into actually using JavaScript. We used an **if** command to display a message when our example dice program threw a double. Let's now look in a bit more detail at **if**, starting with the condition part of the statement.

As a reminder, here is the code that detects a double being thrown, using the two dice scores of **x** and **y**:

```
if (x == y) {
    console.log("DOUBLE");
}
```

The == operator compares two numbers. What is less obvious is that in comparing the numbers, it produces a result that is either **true** or **false**, depending on whether they are equal. A value that can be either **true** or **false** is called a Boolean value. This is a type of data, just like a number or a string.

Try running the following code:

```
var b = (10 > 5);
console.log(b);
```

You will see the value **true** displayed in the console output. This means that the variable **b** has been assigned the value **true**. Change 5 to 50 in this example and run the program again. This time, **b** will have a value of **false**.

In addition to the == operator, which returns **true** if the two values on either side of it are equal, there are these other operators:

- < (less than)
- <= (less than or equal to)
- > (greater than)
- >= (greater than or equal to)
- != (not equal to)

You can think of these comparison operators as being rather like arithmetic operators (such as +), except that instead of producing a number as a result, they produce a Boolean value.

You sometimes need a more complicated condition than a simple comparison of two values. For example, if we wanted to display a message only if double six was thrown, we could write the condition like this:

```
if ((x == 6) && (y == 6)) {
  console.log("Double 6 thrown!");
}
```

There are now two "mini-conditions" on either side of the "&&" operator. The **&&** logical operator is equivalent to the English word "and" and means "and" in a logical sense. That is, the result of the "and" is **true** if, and only if, both sides of the **&&** are **true**.

In addition to **&&**, you can also use **||** (or). The **|** symbol is the "pipe" character on a keyboard, usually found on the backslash key, The "or" operator will return **true** if either of the expressions on either side of it return **true**.

If we wanted to display a special message when the dice throw adds up to 7, 11, or a double, we could write this:

```
if ((total == 7) || (total == 11) || (x == y)) {
   console.log("7, 11 or Double thrown");
}
```

Notice how we have used parentheses to make the meaning of the expression more obvious. That is, we want the **==** comparisons to take place before the **||** (or) operations.

One final logical operator is the "not" operator, which is written as **!**. This applies to a single value and inverts it. So, if the value is **true**, it results in **false**, and vice-versa. Try out the following example to see how this works:

```
var b = true;
console.log(b);
b = ! b;
console.log(b);
```

The console output will display **true** first and then **false** as **b** is toggled by **!**.

Another complexity of the **if** command is that you can also add an **else** statement to allow you to do one thing if a condition is **true** and another if it is **false**. Here's an example:

```
if (x == y) {
    console.log("DOUBLE");
}
else {
    console.log("Not a double");
}
```

You can take this a stage further, adding intermediate **else if** clauses. For example, if we wanted to display different messages depending on the magnitude of a number, but wanted to ensure that only one message was displayed, we could write the following code:

```
var x = 30;

if (x < 10) {
    console.log("x is small");
}
else if ((x >= 10) && (x < 50)) {
    console.log("x is medium");
}
```

```
else {
    console.log("x is large");
}
```

Try changing **x** to a few different values, to make sure that you understand how the **if** and **else** statements are working.

# Functions

Just as a variable is a way of giving a name to a value, a function can be thought of as giving a name to a chunk of code. This is a very useful feature for programming because it makes it easy to use lines of code over and over again.

Many of JavaScript's "commands," such as **random** and **log**, are actually functions that other people have written to make it easy for us to do something.

## A Dice-Throwing Function

Let's start by taking our dice-throwing example and putting some of the code into a function:

```
function throwDice() {
    var r = Math.random() * 6;
    r = Math.floor(r) + 1;
    return r;
}

var x = throwDice();
console.log(x);
var y = throwDice();
console.log(y);
```

The key thing about functions is that the lines of code inside the function are not run until the function is "called." So, whereas in our earlier code, every line of code would be run in turn automatically, now we are giving a name to a chunk of code that we may decide to run later. In other words, we first define a function, specifying the lines of code it contains, and then later we may decide to run or "call" the function, which is when the code inside the function definition will actually run.

In the example, this only happens when we get to the line

```
var x = throwDice();
```

and then again when we get to the following line:

```
var y = throwDice();
```

So placing the word "function" in front of the name of the function indicates that the function is being defined, and using the function name on its own with parentheses after it causes the lines of code in the function to be run.

## Naming Functions

To define a function in JavaScript, we use the command **function**, followed by the name of the function (in this case, **throwDice**). The rules for naming functions are the same as those for naming variables that you learned in Chapter 3. That is, they should begin with a lowercase letter, with each new word starting with an uppercase letter. It is very important that the name of the function describes the purpose of the function. Whereas variable names tend to be nouns or noun phrases, because they describe things ("count" or "numberOfDice"), function names are usually verbs or verb phrases because they describe what the function will do ("log," "throwDice," and so on).

## The Function Body

After the function's name, we have the parentheses. In this case, these are together, but later on, we will see how the function can be given parameters inside those parentheses. The code of the function (called its "body") is then contained between { and }.

The first two lines of the body of the **throwDice** function are listed here:

```
var r = Math.random() * 6;
    r = Math.floor(r) + 1;
```

They are almost the same as when we were experimenting with dice throws without using a function. The first line defines a variable called **r** and assigns it a value using the **Math.random** function. Because the variable definition is inside the function, the variable can only be used inside this function, so you could use the variable name **r** inside other functions and there would not be any conflict or ambiguity about which "r" was being used. The variable **r** is called a "local" variable, because it is local to the function.

The problem with this is that we want to know the value of the dice throw outside of the function so that we can display it using **console.log**. To do this, we use the **return** command:

```
return r;
```

This specifies that the function is to return the value held in the local variable **r**. So, in the same way as the built-in function **Math.random** returns a number greater than 0 and less than 1, our function is going to return a number between 1 and 6. By returning a value, it can be assigned to a variable, so we can write the following:

```
var x = throwDice();
```

## Locals and Globals

The variable **x** is not inside any function, so it is called a "global" variable. Such variables can be seen throughout the whole program. They should be used sparingly because when programs start to become complex, using global variables can make them very difficult to debug.

This concept of local and global variables is called "scoping." It is important to understand this so as to avoid problems. If you find that you are pretty sure you are changing the value of a variable, but when you print its value using **console.log** it appears unchanged, then scoping problems should be top of your list of suspects.

## Function Parameters

The **Math.random** function returns a decimal number greater than 0 and less than 1. This means that if we have an example like the dice example, we have to do a little arithmetic to turn that returned number into an integer between 1 and 6. Because wanting a random integer is a common problem, we could make our own function that uses **Math.random** but gives us back an integer between some range of values. We can specify that range by giving minimum and maximum random integers as "arguments" to the function. So, let's assume we write the following:

```
var x = randomInteger(1, 6);
console.log(x);
```

We would expect this to do the same thing as if we were using the original **throwDice** function, except that if we suddenly decided that we wanted to throw a 12-sided dice, we could just write this:

```
var x = randomInteger(1, 12);
console.log(x);
```

Let's now write the function definition for **randomInteger** and some lines of code to call the function and test it out:

```
function randomInteger(minInt, maxInt) {
    varspan = maxInt - minInt + 1;
    var r = Math.random() * span;
    r = Math.floor(r) + minInt;
    return r;
}

var x = randomInteger(1, 6);
console.log(x);
var y = randomInteger(1, 6);
console.log(y);
```

This functions exactly as the previous example did. The difference is that we now have a general-purpose random integer-generating function rather than one that only generates a number between 1 and 6.

To prove this, try modifying one of the test lines at the end to extend the range between 1 and 12:

```
var y = randomInteger(1, 12);
console.log(y);
```

Looking at the **randomInteger** function, you can see that now we have something inside the parentheses on the first line of the definition:

```
functionrandomInteger(minInt, maxInt) {
```

These things that look like variable names are called "parameters." They are how we get values into the function so that it can use them; they are just like local variables. Therefore, we can use them within the function. The first line of the function body now calculates the span of numbers between the minimum random number and the maximum. In the case of regular six-sided dice, the "span" is $6 - 1 + 1 = 6$. This then gets used in the next line:

```
var r = Math.random() * span;
```

The parameter "**minInt**" is then used again to add an offset to the random number range.

The math behind this is not too important, but the idea that you use parameters to get values into a function and then use **return** to specify an output from the function is an important concept.

# While Loops

In Chapter 3 we looked at the **for** loop and discovered how to use it to run some commands a fixed number of times. If you don't remember how this works, have a quick look at the "Looping" section of Chapter 3 before continuing.

Sometimes, you'll want to keep doing something while some condition is true. As an example, let's say we want our program to keep throwing dice until it gets a double. First, let's tidy the dice-throwing program up a bit.

On the book's web page (www.beaglebonebook.com) you will find a link to the program files on github. Each program in its own file, identified by the chapter number and then the listing number (in this case, "04" for chapter 4 and "01" for the first listing). You can just browse these files in github and then copy and paste the text of the programs into an empty editor window on Cloud9. However, a more convenient approach is to download all the example programs for the book and install them directly into Cloud 9's area for programs. To do this, follow these steps:

1. Open an SSH session or terminal window on the BeagleBone Black.

2. Change the working directory to Cloud9's program folder using the following command:

   ```
   # cd/var/lib/cloud9
   ```

3. Fetch the example programs from github using this command:

   ```
   # git clone git://github.com/simonmonk/prog_bbb.git
   ```

4. This fetches the files but places them inside an extra directory that we do not need here, so enter the following command:

   ```
   # cp "prog_bbb/Prog BBB/"* "Prog BBB"
   # rm -fr prog_bbb/
   ```

*NOTE*   *If you already have a folder called "Prog BBB", its contents will be overwritten, so you may wish to rename this folder.*

5. Delete the unwanted extra folder using this command:

   ```
   # rm -r prog_bbb/
   ```

6.  Refresh Cloud 9 in your browser, and the new folder full of programs should now appear in the projects area (see Figure 4-1).

Try running the **04_01_dice_simple.js** program and you will see that the output is a bit more readable. We have formatted it all onto one line using + to join together the strings and numbers into a single message to be displayed:

```
Throw: 2, 1
```

If we wanted to change the program so that it just keeps rolling the dice until a double-six is thrown, we could use a **while** loop like this:

```
var x = throwDice();
var y = throwDice();
console.log("Throw: " + x + ", " + y);
while ((x != 6) || (y != 6)) {
    x = throwDice();
    y = throwDice();
    console.log("Throw: " + x + ", " + y);
}
```

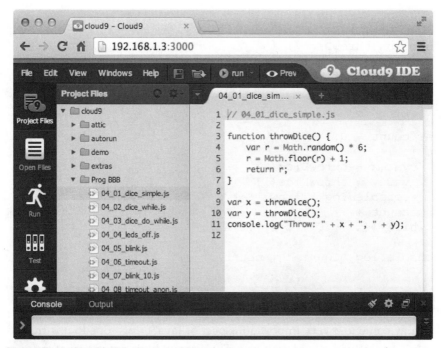

**Figure 4-1**  *Cloud9 IDE after installing the example programs*

You can find the full program in the file 04_02_dice_do_while.js. When you run it, you are likely to see lots of throws of the dice ending with a double-six.

Looking at the code, you can see after two dice are thrown, the result is displayed and then we have the start of a **while** loop. Rather like the **if** command, **while** is followed by a condition. The condition is for staying in the loop. The program will keep running the lines of code inside the { and } as long as **x** and **y** are not both sixes. The code inside the loop is a repeat of the three lines before the **while** that throw the dice again and display the result.

Repeating lines of code in a program is often a bad sign, indicating that the program could be made simpler. It usually makes sense to place repeated code into a function, but in this case we can use a different form of the **while** loop that makes the test as to whether to stay in the loop at the end rather than at the beginning of the loop. That way, the dice will have already been thrown once before we get to the condition.

You can find the code for this in the file 04_03_dice_do_while.js. We have also added the refinement of counting how many rolls it takes to get a double-six using the variable **count**:

```
// 04_03_dice_do_while.js

function throwDice() {
    var r = Math.random() * 6;
    r = Math.floor(r) + 1;
    return r;
}

var count = 0;
do {
    var x = throwDice();
    var y = throwDice();
    console.log("Throw: " + x + ", " + y);
    count ++;
} while ((x != 6) || (y != 6));

console.log("Count=" + count);
```

# Timers

We now come to a very important aspect of JavaScript—one that is treated quite differently in other languages. The concept boils down to making sure that things do not hang around waiting when there is nothing to do. It sounds like something simple and also like something that should be the responsibility of the operating system rather than you as a programmer, but JavaScript takes an interesting approach to this problem.

## No Sleeping

Most languages have a built-in command called **sleep** that takes a parameter indicating the amount of time to sleep. Then, the program using **sleep** tells the operating system that it has nothing to do for the amount of time indicated. The operating system will usually take this hint and use the time productively to do other things such as update the position of your mouse or give attention to any one of the other dozens of processes that vie for its time on a Linux computer. After the allotted time, the operating system remembers to return to the program with the **sleep** command in it so that it continues on after sleeping. For example, the following example is written in Python and taken from a program that flashes an LED on and off once per second on a Raspberry Pi, so do not try and run it on your BeagleBone Black.

```
while (True):
GPIO.output(18, True)
time.sleep(0.5)
GPIO.output(18, False)
time.sleep(0.5)
```

So, pin 18 is turned on, we wait for half a second, the pin is turned off again, and we wait for another half second, and so on. During those two half-second pauses, the operating system is free to go off and be productive.

JavaScript was originally designed to be run in a browser as part of what the browser is doing (that is, mostly drawing web pages on your computer screen), so it does not really participate in the operating system's scheduling other than as part of the browser's process. This means that if you had a loop like the preceding Python code, the browser would appear to hang.

## Interval Timers

JavaScript's solution to this problem is to encourage (very strongly) writers of JavaScript to always write functions that do whatever they are doing and then immediately return. They should not be hanging about waiting for anything. This is backed up by a "timer" mechanism that schedules a function to be called at regular intervals.

Before we look at how we would make an LED blink using a timer, let's try out a really simple example that will just write something to the console output. Try running the following small program in your play.js file:

```
function sayHello() {
    console.log("Hello World");
}

setInterval(sayHello, 1000);
```

This first defines a function called **sayHello** and then uses **setInterval** to ensure that it is called every 1,000 milliseconds (every second). This will continue writing out the words "Hello World" until you click the red "stop" square in Cloud9. The **setInterval** function takes two parameters. The first is a function name (in this case, **sayHello**) and the second is a time interval in milliseconds.

## Blinking an LED with set Interval

You will need to see what the USR LEDs are doing, so let's turn them all off using the following short program (04_04_leds_off.js):

```
// 04_04_leds_off.js

var b = require('bonescript');
for (var i = 0; i< 4; i++) {
    var pin = "USR" + i;
    b.pinMode(pin, b.OUTPUT);
    b.digitalWrite(pin, 0);
}
```

This program is a bit of a diversion, but it's a useful way of turning off all the LEDs so that we can get a better view when we make the LED blink. Notice how it uses a **for** loop to construct a name for "pin" for each of the four "USR" LEDs. You will learn more about this topic in the next chapter.

Now, let's write a program to make the LED blink. This program is called 04_05_blink.js:

```
// 04_05_blink.js

var b = require('bonescript');

var led = "USR3";
b.pinMode(led, b.OUTPUT);
```

```
var state = 0;

function toggleLED() {
    state = state ^ 1;
    b.digitalWrite(led, state);
}

setInterval(toggleLED, 500);
```

Run it, and you will see LED USR3 blink, once per second. Let's now examine the code more closely.

The first line imports BoneScript so that we can have access to the LEDs. The next line specifies which LED we are going to blink, and the line after that defines it as being an OUTPUT.

We need a global variable in which to record whether the LED is on or off. The variable is called **state** and has a value initialized to be 0 (off). The function definition uses the ^ **1** command to toggle **state** between 0 and 1. The second line of the **toggleLED** function then sets the LED to that state using **digitalWrite**.

The **timeInterval** function is then used to run **toggleLED** every half-second.

## setTimeout

The function **setInterval** is fine for a repeated event, but what if you just need to do something once after a certain time period, or want to vary the time period between each run of the function? In these cases, you can use **setTimeout**.

Try running the following example (04_06_timeout.js):

```
// 04_06_timeout.js

console.log("Hello");

function sayMore() {
    console.log("World");
}

setTimeout(sayMore, 5000);
```

This will write "Hello" to the console output and then wait five seconds before writing "World" to the console as well.

So, the difference between **setInterval** and **setTimeout** is that **setTimeout** only happens once, whereas **setInterval** continues until you tell it to stop.

## Cancelling an Interval Timer

If you cast your mind back to our blinking LED, you will remember that it just carries on blinking indefinitely. If we wanted to have the LED blink for 10 seconds and then stop, we could use **setTimeout** to cancel the timer. To be able to know which interval timer to cancel (if there were more than one running), we would need to have a handle on it. This is done by assigning the result of **setInterval** to the variable called **timer**, as shown in the program 04_07_blink_10.js:

```
// 04_07_blink_10.js

var b = require('bonescript');

var led = "USR3";
b.pinMode(led, b.OUTPUT);

var state = 0;

function toggleLED() {
    state = state ^ 1;
    b.digitalWrite(led, state);
}

var timer = setInterval(toggleLED, 500);

function stopTimer() {
    clearInterval(timer);
}

setTimeout(stopTimer, 10000);
```

A timeout is then set to run **stopTimer** after 10 seconds. The **stopTimer** uses the command **clearInterval** to stop the interval timer running.

## Anonymous Functions

Sometimes, you can simplify your code by defining a function within the place where you are going to use it, in which case it does not need a name. To illustrate this idea, let's return to the program 04_06_timeout.js:

```
// 04_06_timeout.js

console.log("Hello");
```

```
function sayMore() {
    console.log("World");
}

setTimeout(sayMore, 5000);
```

This code seems kind of verbose just to write something to the console, wait five seconds, and then write something else. This is because we had to write a whole function called **sayMore** just to write the second part of the message.

We can simplify program 04_06_timeout.js to the following two lines of code:

```
//04_08_timeout_anon.js
```

```
console.log("Hello");
setTimeout(function(){console.log("World");}, 5000);
```

At first sight, the second line looks like a horrendous mixture of punctuation and brackets. Figure 4-2 shows what the various parts of this line do.

The first parameter to **setTimeout** is not now the name of a function, but actually a whole function. Because the function is only used here, it does not need a name, so none has been given. Incidentally, the program would still work if you did give it a name; there just isn't any need for one.

Figure 4-2 shows the exploded view of the function. This has been reformatted as you would normally write it. Note that only spaces and new lines

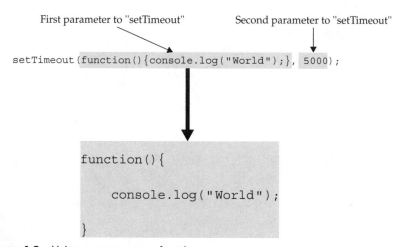

**Figure 4-2**  *Using an anonymous function*

have been added; the actual code of the function is the same as the unexploded version.

Using functions like this is pretty common in JavaScript. These type of functions are called "anonymous functions." You will also sometimes hear them called "lambda functions," which sounds altogether much more impressive.

## Summary

In this chapter we explored the key JavaScript features of functions and timers. You also learned a bit more about **if** statements and loops, using what you have learned to make an LED blink.

In the next chapter, we will build on this and look at the key data structures of JavaScript called arrays and objects. We will also create a blinking LED example to use Morse code.

# 5

# Arrays, Objects, and Modules

In this chapter you will learn how to use JavaScript arrays, and we illustrate their use with BoneScript to make Morse code signals using the built-in LEDs. You will also be introduced to JavaScript's concept of "objects," which provides a very powerful but unusual mechanism for object-oriented programming.

## Arrays

The variables you have encountered so far have only had a single value. JavaScript arrays are used to contain a list of values. You can think of an array as being rather like a stack of pigeonholes. You can take something out of a particular location, using its index position. In this case, 0 is used to refer to the first position, 1 to the next, and so on.

## Creating Arrays

The following line of JavaScript will create an array containing five numbers, but array slots can contain any type of data, including strings and even other arrays:

```
var numbers = [123, 34, 55, 321, 9];
```

Figure 5-1 shows this in diagrammatic form.

Arrays in JavaScript work a little differently than arrays or lists in other languages, so to get a feel for how arrays work, open a terminal session on

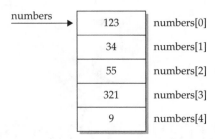

**Figure 5-1** *A JavaScript array*

your BeagleBone Black and run Node. You can then play with arrays in the JavaScript command line (see Chapter 1). In the following transcript, the > prompt indicates the Node prompt where you can type a line of code, and Node's response will then follow on the subsequent lines. Note that, when using the Node command line, you can just enter the name of a variable on its own, and Node will display is value without having to use console.log.

You can also omit the semicolon on the end of the line because we are only entering one line at a time.

```
>var numbers = [123, 34, 55, 321, 9];
undefined
> numbers
[ 123, 34, 55, 321, 9 ]
>
```

Here, we have created a variable called **numbers**, containing the five numbers as shown. This is confirmed when we enter **numbers** on the Node command line and the contents of the array are displayed for us to see.

## Accessing Elements of an Array

We can now try accessing the first thing in the array using [ and ] around the index position of the item we want:

```
>numbers[0]
123
>
```

Similarly, we can get the second item (or "element" as they are normally called):

```
>numbers[1]
34
```

We can also change elements of the array like this:

```
>numbers[4] = 444;
444
> numbers
[ 123, 34, 55, 321, 444 ]
>
```

You can see that the contents of the last element of the array has been changed to 444.

We can find the length of the array like this:

```
>numbers.length
5
>
```

If you try to access an element of the array that does not exist because the array index is too high, you will get a result like this:

```
>numbers[100]
undefined
>
```

You can use this special value of **undefined** in conditions. For example, you could write the following:

```
x = numbers[100];
if (x == undefined) {
  console.log("x is undefined");
}
```

## Modifying Arrays

You can add more elements to the end of an array using **push**:

```
>numbers.push(2);
6
> numbers
[ 123, 34, 55, 321, 444, 2 ]
>
```

The value returned by **push** is the number of elements in the array.

Most array manipulations (both adding and removing items) can be accomplished using **splice**. The first parameter of **splice** is the index position where the splicing is to take place. The second parameter is the number of items to be removed. If you are using **splice** to add things rather than remove anything, this will be 0. You can then follow this parameter with as many parameters as you like, and each of the values in those parameters will be added into the array.

This example creates a new array (x) and removes the second and third elements from it:

```
>x = [10, 20, 30, 40, 50]
>x.splice(1, 2)
[ 20, 30 ]
> x
[ 10, 40, 50 ]
>
```

There are two interesting things to note here. First, the result of the splice is a list containing the items that have been removed (which could be useful). Second, when we look at the contents of the list **x**, we can see that the two items have been removed.

Now let's use **splice** to put back the two elements we have just removed:

```
>x.splice(1, 0, 20, 30)
[]
> x
[ 10, 20, 30, 40, 50 ]
>
```

In this case, we want to insert the new elements at index position 1, so the first parameter will be **1**. There are no elements to be deleted, so the second parameter will be **0**. The other two parameters are the items to be added (**20** and **30**). With them added, when we enter just **x** onto the command line, we can see that the array is now fully restored to its original value.

Although these examples have used numbers as the values in the array, there is no reason why the elements couldn't be strings or Booleans or even other arrays.

## Sorting Arrays

You can use **sort** to put the elements of an array into order. Here's an example:

```
>x = [7, 2, 6, 0, 1]
[ 7, 2, 6, 0, 1 ]
>x.sort()
[ 0, 1, 2, 6, 7 ]
```

If we want them to be sorted in descending order rather than ascending order, we need to supply **sort** with a function with which to compare individual elements. This can be an anonymous function. Try out the following example:

```
>x = [7, 2, 6, 0, 1]
[ 7, 2, 6, 0, 1 ]
>x.sort(function(a, b) { return (a < b);})
[ 7, 6, 2, 1, 0 ]
>
```

The way to think of this is that any kind of sorting method will ultimately have to compare two of the things it is sorting, to decide on their order in the array. By doing enough of these in an organized way, it will be able to put the entire array into the order specified by the comparison.

## Iterating Over Arrays

In programming, the word *iterating* means repeating something a number of times. This is just what we were doing with the **for** and **while** loops. In the case of arrays, you may want to "iterate over" an array, which means looking at each element of an array and possibly doing something with it.

We can do this using a simple **for** loop. Try out this example in play.js:

```
var people = ["fred", "mary", "becky", "steve"];

for (var i = 0; i < people.length; i++) {
    console.log(people[i]);
}
```

This approach uses an index variable (**i**) inside the **for** loop access the appropriate element in turn using the [] notation.

Because iterating over arrays is such a common thing to do, there is an alternative way of doing this that is usually more concise. This way, shown here, applies an anonymous function to each of the elements of the array in turn:

```
var people = ["fred", "mary", "becky", "steve"];

people.forEach(function(person) {console.log(person);});
```

The anonymous function that is the parameter to **forEach** itself has one parameter (**person**) that will be assigned to each of the values in the array in turn. The body of this function simply writes the element to the console output.

Appendix B contains a section on arrays where you can find out about the commands available.

# Morse Code Example

In this example program we use arrays to keep everything concise and easy to read. Let's start with a simple modification to the blink example from Chapter 4 (04_05_blink.js). Because we are going to develop an example that flashes Morse code, let's at least have all four built-in LEDs flash in unison rather than just one solitary LED.

Morse code is designed to send letters and digits from the Roman alphabet. Each letter is made up of dots and dashes. For example, the letter *A* is dot-dash (.-) in Morse code. The resulting code can be sent as a series of long or short tones, or as long or short flashes of light. We are going to use the built-in LEDs to flash Morse code as light.

You can find out more about Morse code at http://en.wikipedia.org/wiki/Morse_code.

## Flashing All Four LEDs at Once

Start by loading and running 04_05_blink.js from Chapter 4:

```
var b = require('bonescript');

var led = "USR3";
b.pinMode(led, b.OUTPUT);

var state = 0;

function toggleLED() {
    state = state ^ 1;
    b.digitalWrite(led, state);
}

setInterval(toggleLED, 500);
```

Instead of using a regular variable to identify the name of the LED, let's use an array. Therefore, we can replace the line

```
var led = "USR3";
```

with

```
varleds = ["USR0", "USR1", "USR2", "USR3"];
```

To initialize all the LED pins as outputs, we can then write the following:

```
leds.forEach(function(led) { b.pinMode(led, b.OUTPUT); });
```

Finally, to toggle the LEDs on and off, we can change the line

```
b.digitalWrite(led, state);
```

to

```
leds.forEach(function(led) { b.digitalWrite(led, state); });
```

Here is the full listing to blink all the LEDs at once. Try it out:

```
// 05_01_blink_all_leds.js

var b = require('bonescript');

varleds = ["USR0", "USR1", "USR2", "USR3"];
leds.forEach(function(led) { b.pinMode(led, b.OUTPUT); });

var state = 0;

function toggleLED() {
    state = state ^ 1;
    leds.forEach(function(led) { b.digitalWrite(led, state); });
}

setInterval(toggleLED, 500);
```

# Blinking SOS

The internationally recognized distress signal of SOS is sent in Morse code as its three letters. The letter *S* is ... and the letter *O* is ---. Some of the other rules of Morse code are that you can send pulses as long or short, as you like, but the flash for a dash should be three times that of a dot. There should also be a gap of one dot's worth of time between each dash or dot. Between each letter, there should be a gap of one-dash duration.

You can find the full listing for blinking SOS in program 05_02_blink_sos.js. You will find it useful to load this up in Cloud9, try it out, and then follow the description of how it works.

First off, let's define some pulse durations:

```
var dot = 200;
var dash = dot * 3;
var gap = dot;
```

These durations are in milliseconds. By defining **dash** and **gap** in terms of **dot**, we only need to change the value of **dot** to speed up or slow down our Morse code being sent.

The next section of the code initializes the LEDs as outputs, in the same way we did in the previous program. After this, we have two very similar functions, **ledsOn** and **ledsOff**:

```
functionledsOn() { // turn the four on-board LEDs on
    leds.forEach(function(led)
        { b.digitalWrite(led, 1); });
}
```

As you might expect, one turns all the LEDS on and the other turns them all off. These functions are used by the **flash** function:

```
function flash(period) {
    ledsOn();
    setTimeout(ledsOff, period);
}
```

This function takes a duration (of the flash) as its parameter. It turns the LEDs on and then sets a timeout to turn them off again after the specified period.

Most of the action for generating the flashes happens in the function **flashMessage**. This function expects an array of durations as its parameter. To get the timing right for each of the flashes, it uses a variable called **timeline**:

```
functionflashMessage(durations) {
    var timeline = 0;
    for (vari = 0; i<durations.length; i++) {
        var d = durations[i];
        setTimeout(function() { flash(d); }, timeline);
            timeline = timeline + d + gap;
    }
}
```

After each of the durations has been iterated over, a timeout is set with an anonymous function that will run **flash** at the appropriate time. The appropriate time is maintained in the **timeline** variable, to which the duration and a gap after each dot or dash is added.

This example is not perfect. For instance, it does not handle the gaps between letters correctly. It's also not a very convenient way to represent the text to be sent as Morse code. We will be returning to this example a bit later in this chapter, to make it capable of translating any text we should wish to be displayed into Morse code. To understand this upcoming example, you need to learn a bit more about strings and objects.

# More on Strings

Strings in JavaScript are treated a little like arrays, where each character of the string is like an element in an array. This means you can access individual characters of a string like this:

```
> s = 'abcdef';
'abcdef'
> s[1]
'b'
>
```

You can also find the length of a string in the same way as an array:

```
> s.length
6
>
```

Appendix B contains a section on strings, where you can find out about some other commands available.

# Introducing JavaScript Objects

We now come to the most powerful and useful concept in the JavaScript language: JavaScript's take on the concept of an object. If you are used to a programming language where you define classes, create instances of those classes, and then call the objects "instances," then you need to clear your mind of all that. Things are different in JavaScript.

If, on the other hand, JavaScript is your first object-oriented programming language, then rejoice! You have nothing to unlearn.

## Comparing JavaScript to Other Languages

Programming languages can be divided into two main categories. In strongly typed languages, you specify exactly the type of everything before you start using it. So, if a variable is going to be used to contain a string, you must declare it as such. If, on the other land, it is going to contain a number, you must specify that. This makes the language very rigid and precise. It also means that most mistyping or other errors in your program will be flagged before the program is actually run.

On the other hand, weakly typed languages are much more flexible. They give the programmer a great deal more freedom, which also, of course, means the freedom to do really stupid or obscure things. This means that writing good code in a weakly typed language requires caution and common sense on the part of the programmer.

JavaScript is a weakly typed language. As such, it has much more in common with languages such as Python, Ruby, LISP, and Smalltalk than it does with strongly typed languages such as Java, C++, and C#.

JavaScript uses "objects" in an incredibly powerful way, combining what are a number of different constructs in other programming languages into a single very powerful mechanism. The first aspect of JavaScript objects that we will look at is their use as a way for structuring data that our program needs.

So far, we have used simple variables that just contain a number, string, and arrays that contain a list of values. Using an array, we can retrieve a value from it using the index position of the element that we want to retrieve. But what if we need some other way of looking things up in a data structure?

For example, each letter of Morse code has a corresponding sequence of dots and dashes. For example, the letter *A* is the sequence .- (dot-dash). If we are going to make our Morse code flashing program general purpose, we need a way of converting any letter, A to Z, into a sequence of dots and dashes.

We could do this with a whole load of **if** statements, a bit like this:

```
if (letter == 'A') {
    flash('.-');
}
else if (letter == 'B' ) {
  flash('-…');
}
```

Here, **flash** is a function that we have just invented that would flash out the sequence of dots and dashes.

The problem with this approach is that it is just very wordy. Each letter requires four lines of code. It's also inefficient, because if we wanted to flash Z, we would have to step through 25 **if**s that would fail before we came to the comparison for Z. We would not notice the performance as being bad, but this approach lacks elegance.

JavaScript objects can be used as an "associative" data store. That is, a value is associated with a key. So, we can associate the value .- with a "key" of A. When we need a value back out of this associative store, we can just ask for it using the key for the value we want. In other programming languages, this type of data structure may be called a hash table, dictionary (Python), or table.

Start up Node and try the following example:

```
>var obj = {}
undefined
```

This first line defines a new variable called **obj** and gives it an initial value of {}, which means an empty object with no data in it.

You can now put some data into it. Let's add the Morse codes for the letters *A* to *C*:

```
>obj['a'] = '.-'
'.-'
>obj['b'] = '-...'
'-...'
>obj['c'] = '-.-.'
'-.-.'
>
```

We can now see the contents of **obj**:

```
>obj
{ a: '.-', b: '-...', c: '-.-.' }
>
```

Just as with an array, when we want to fetch a value out of the data structure, we use a square bracket. However, whereas when using an array we would put an index value inside the square brackets, when using an object, we put the key. Here's an example:

```
>obj['a']
'.-'
>
```

The value that we store against a key is a string in this case, but it does not have to be. It can be of any type. It could be a number, an array, or other object.

It can help to visualize what is going on with an associative store like this. Figure 5-2 shows the use of an object as an associative store.

A nicety of JavaScript is that we can also access elements of the object using a dot notation:

```
>obj.a
'.-'
>
```

At this point, if you have programmed in other object-oriented languages, you are allowed to gasp in awe at the cleverness of this approach.

Just as with arrays, you can also initialize an object with values. However, each item that you are adding to the object must supply a key and a value and so the syntax is a bit more complex. So, initializing **obj** with the letters *A* to *C* would look like this:

```
>obj = { "a": '.-', "b": '-...', "c": '-.-.' }
{ a: '.-', b: '-...', c: '-.-.' }
>
```

If the keys used are single words, with no spaces (like a variable name), you do not need to put them in quotes. Therefore, we can also initialize the object like this:

```
>obj = { a: '.-', b: '-...', c: '-.-.' }
{ a: '.-', b: '-...', c: '-.-.' }
>
```

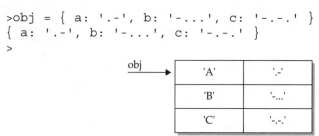

**Figure 5-2** *An object*

# Morse Revisited

You now know enough about objects to continue with our Morse code example and make it more general purpose so that we can have it flash any message we like. You can find the full code for this example in program 05_03_blink_morse.js. You will find it useful to load this up into Cloud9 as the program is explained. You should also try running it.

The first four lines of code are the same as the previous program (05_02_blink_sos.js). The BoneScript library is imported and variables defined for the flash durations. Next, we define an object called **letters** and initialize it with all the Morse codes for the letters *a* to *z*:

```
var letters = {
    "a" : ".-",
    "b" : "-...",
    "c" : "-.-.",
// etc.
    "z" : "--..",
    " " : "   "
};
```

Note that we have also added a value of three spaces associated with a key of a single space. We will use this later to cause the correct gap to appear between words being flashed.

The LEDs are initialized in exactly the same way as the previous program. The **ledsOn, ledsOff,** and **flash** functions stay the same as well:

```
varleds = ["USR0", "USR1", "USR2", "USR3"];
leds.forEach(function(led) { b.pinMode(led, b.OUTPUT); });


function ledsOn() { // turn the four on-board LEDs on
    leds.forEach(function(led) { b.digitalWrite(led, 1); });
}

function ledsOff() {
    leds.forEach(function(led) { b.digitalWrite(led, 0); });
}

function flash(period) {
    ledsOn();
    setTimeout(function() { ledsOff(); }, period);
}
```

The **flashMessage** function, however, is quite different. It takes as its parameter a string that consists of characters that are either a dot, dash, or space. As before, the **timeline** variable is used to keep track of the next time that the LED needs to flash, but now as we loop over each dot, dash, or space in the message, three **if** clauses handle the situations for these characters separately.

```
function flashMessage(message) {
    var timeline = 0;
  for (var i = 0; i < message.length; i++) {
    if (message[i] == '.') {
        setTimeout(function() { flash(dot); }, timeline);
        timeline = timeline+ dot + gap;
    }
    else if (message[i] == '-') {
      setTimeout(function() { flash(dash); }, timeline);
        timeline = timeline + dash + gap;
    }
    else if (message[i] == ' ') {
        timeline = timeline +  dot + gap;
    }
   }
  }
}
```

If the element of the message is a dot, then a flash of the appropriate duration is made and the timeline updated. The handling of a dash is similar. If a space is encountered, there is nothing to flash, but we do want to move the **timeline** along for the next dot or dash.

The function **flashText** takes an arbitrary text message (as long as it only uses the letters *a* to *z* and spaces) as its parameter. It then goes through each character in turn, converting it into dots, dashes, and spaces, and concatenates it to the variable **dotsAndDashes**. When it has assembled all the characters, it logs them to the console and then calls **flashMessage** to actually cause the LEDs to flash.

```
function flashText(text) {
  var dotsAndDashes = "";
  text = text.toLowerCase();
  for (var i = 0; i<text.length; i++) {
    dotsAndDashes += letters[text[i]] + " ";
  }
  console.log(dotsAndDashes);
  flashMessage(dotsAndDashes);
}
```

One neat trick is that to avoid problems with the message containing upper and lowercase letters, each letter is converted into lowercase using **toLowerCase()**.

Finally, to flash the message, we can just call **flashText**, like this:

```
flashText("It was the best of times");
```

Spend some time with this program, changing the message. Find out what happens if you put an unexpected character in the message (a number, for example). Try modifying the program so that it also flashes the digits 0 to 9.

# Debugging JavaScript

The word "debugging" means fixing problems with code that is not working right. Software will occasionally behave in a manner that is unexpected and baffling. Tracking down exactly what is going wrong can be a tricky business.

Cloud9 has built-in features to help you with debugging. In particular, you can set *breakpoints* in the code, where the program will stop running so you can take a look at the variables and then step through the code a line at a time.

To enable debugging for a particular program, you need to click the drop-down arrow right next to the green Run button and then select Run in Debug Mode. This will change the Run button's label to Debug.

You can now set a breakpoint in your code by clicking in the margin to the left of the line of code where you want the breakpoint. When you do this, a yellow dot will appear. This is shown in Figure 5-3, which displays the program 04_03_dice_do_while.js being debugged.

When you run the program by clicking the Debug button, it will stop at the breakpoint line. You can now control the progress of the program by clicking the controls that appear at the top-right of the code window. Use the hover help functionality to see what each of these does.

The first triangular Play-type button resumes the program, which will then continue until it hits another breakpoint. The Step Over button will step over the current line, go on to the next line, and then stop again. The less frequently used Step Into button will, if the line has a function such as **while** in it, take you into that function rather than jump over it. Its counterpart, Step Out Of, will finish that function and return to where you were.

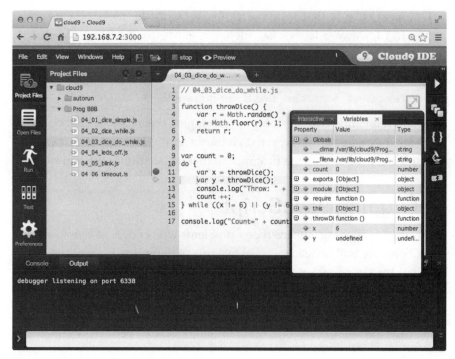

**Figure 5-3**  *Debugging in Cloud9*

You can see the values of the variables in Figure 5-3. In this case, **x** has just been assigned the value 6 as a result of the previous line. To inspect the variables, click the icon that looks like a microscope.

# Modules

Modules provide a way of grouping together chunks of code so that you can use them from a number of JavaScript programs. We have already used the **bonescript** module in our programs by adding the following line to the top of them:

```
var b = require('bonescript');
```

We can then use functions defined in **bonescript** by placing **b.** in front of them, like this:

```
b.digitalWrite(led, 1);
```

As well as using modules that other people have written, it is a good idea to write code that may be suitable for reusing as modules. As an example, we could convert our final Morse code program (05_03_blink_morse.js) to be a module. Having done so, we could then just use it like this:

```
var m = require('./morse');
```

```
m.flashText("It was the best of times");
```

The **require** statement names the module to use **./morse**. The "**./**" indicates that the module code will be contained in the same directory as the program running it. Also, the name of the module does not have ".js" on the end because this is assumed.

To convert 05_03_blink_morse.js into a module in the file morse.js (included in the downloads), just three lines of code need to be changed. You can also find a test program that uses the module in 05_04_morse_using_module.js.

First, any functions in the module that we want to be accessible to programs using the module must be "exported." This requires the first line of the function (in this case **flashMessage**) to be changed from

```
functionflashMessage(message) {
```

to

```
exports.flashMessage = function(message) {
```

The name of the function has been moved so that it is now a variable, prefixed with **exports.** This is then assigned to the now anonymous function. The same change needs to be made to the function **flashText**.

Within the body of the function **flashText** is a reference to **flashMessage**, which must be altered to **exports.flashMessage**, as highlighted here:

```
exports.flashText = function(text) {
  var dotsAndDashes = "";
  text = text.toLowerCase();
  for (vari = 0; i<text.length; i++) {
    dotsAndDashes += letters[text[i]] + " ";
  }
  console.log(dotsAndDashes);
  exports.flashMessage(dotsAndDashes);
}
```

The module should not actually be flashing a message without some other program telling it to do so. Therefore, we should also remove the final line of the program:

```
flashText("It was the best of times");
```

When you are writing a program, it is a good idea to consider whether parts of your program should actually be written as separate modules. This helps to keep the complexity of the programs manageable as well as makes it possible for others to take and use the module you created.

## Summary

In this chapter you learned some fairly advanced JavaScript. In the next chapter we look at the business of using BoneScript to control the GPIO pins and finally start experimenting with a bit of electronics.

# 6

# BoneScript

You could use the BeagleBone Black and JavaScript without ever using the GPIO connections, but you would miss out on the full BeagleBone Black experience. In this chapter, you will get to grips with the nuts and bolts of the GPIO connectors and see how to use BoneScript to read and write to the GPIO pins.

BoneScript is the JavaScript module that allows us to control the GPIO pins and the built-in LEDs of the BeagleBone Black. In Chapter 4 you used this library to control the built-in LEDs by using **digitalWrite**. You can also read digital inputs, read analog inputs, and write analog outputs.

## GPIO Connectors

The BeagleBone Black has a large number of GPIO (General Purpose Input/ Output) connections available in the form of two long thin sockets on either side of the board. You can use these for attaching external electronics in the form of plug-in "capes" that fit over the BeagleBone Black (see Figure 6-1). Alternatively, you can link the sockets via jumper wires to a solderless bread-board on which other components can be placed (see Figure 6-2). You will meet the solderless breadboard properly in Chapter 7, where you will also have the chance to make the arrangement shown in Figure 6-2, which controls the speed of a motor using a BeagleBone Black.

The connections available on the two sockets are summarized in Figure 6-3. You can also find this diagram in Appendix C. It is probably something you will frequently need to refer to because the socket connections are not heavily labeled on the board itself. One way around this is to stick paper labels to the

81

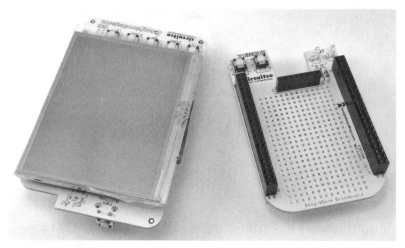

**Figure 6-1** *LCD3 and breadboard capes*

side of the GPIO sockets. The BeagleBone Black in Figure 6-2 has ready made templates from monkmakes.com attached to it.

Note that even though the connections of the sockets are female rather than male, they are still often referred to as "pins."

All the pins labeled P8_NN or P9_NN, where NN is the number on the connector (P8 or P9), can be used as both digital outputs and digital inputs. Those pins, also marked "PWM," are capable of analog output. Some of the pins also have "RX" or "TX" followed by a number after them. These pins can also be used for serial interfacing.

**Figure 6-2** *A BeagleBone Black linked to breadboard*

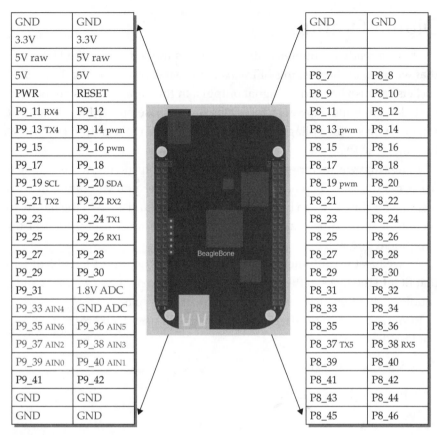

| | | | | |
|---|---|---|---|---|
| GND | GND | | GND | GND |
| 3.3V | 3.3V | | | |
| 5V raw | 5V raw | | | |
| 5V | 5V | | P8_7 | P8_8 |
| PWR | RESET | | P8_9 | P8_10 |
| P9_11 RX4 | P9_12 | | P8_11 | P8_12 |
| P9_13 TX4 | P9_14 pwm | | P8_13 pwm | P8_14 |
| P9_15 | P9_16 pwm | | P8_15 | P8_16 |
| P9_17 | P9_18 | | P8_17 | P8_18 |
| P9_19 SCL | P9_20 SDA | | P8_19 pwm | P8_20 |
| P9_21 TX2 | P9_22 RX2 | | P8_21 | P8_22 |
| P9_23 | P9_24 TX1 | | P8_23 | P8_24 |
| P9_25 | P9_26 RX1 | | P8_25 | P8_26 |
| P9_27 | P9_28 | | P8_27 | P8_28 |
| P9_29 | P9_30 | | P8_29 | P8_30 |
| P9_31 | 1.8V ADC | | P8_31 | P8_32 |
| P9_33 AIN4 | GND ADC | | P8_33 | P8_34 |
| P9_35 AIN6 | P9_36 AIN5 | | P8_35 | P8_36 |
| P9_37 AIN2 | P9_38 AIN3 | | P8_37 TX5 | P8_38 RX5 |
| P9_39 AIN0 | P9_40 AIN1 | | P8_39 | P8_40 |
| P9_41 | P9_42 | | P8_41 | P8_42 |
| GND | GND | | P8_43 | P8_44 |
| GND | GND | | P8_45 | P8_46 |

**Figure 6-3**　*The BeagleBone Black GPIO connectors*

You'll notice some gaps in the pin labels. These pins are not currently used, but may be in later versions of the board.

GND (0V) connections are available on both P8 and P9, and P9 also has 3.3 and 5V connections. The 5V connection marked "RAW" is connected directly to the 5V barrel jack. It will not supply any voltage when the board is only powered over USB. The pins marked 5V will always supply 5V while the BeagleBone Black is powered up, whether it is through the DC jack or over USB.

P9 has a section of pins for analog-to-digital conversion (ADC). There are seven analog inputs from which voltages up to 1.8V can be measured. The ADC also has its own two power pins (labeled "GND_ADC" and "1.8V ADC"). These should be used as the reference voltages when you're connecting analog electronics because they are designed to be "low noise," thus leading to more accurate analog readings.

## Digital Outputs

We have already encountered digital outputs in the form of the USR LEDs that we used in the flashing LEDs example. Most of the pins on both GPIO sockets can also be used as digital outputs. In this case, the pins are not connected to anything. However, by setting one of these pins to be a digital output of 1, you set the voltage on the pin to 3.3V. Setting the output to 0 will set the voltage to 0V.

To test that we are indeed turning the outputs on and off, we are going to use a multimeter. The cheapest digital multimeter you can buy will work just fine for this. You can buy these at RadioShack and even in the hardware section of some supermarkets. This is the single most useful item of test equipment you will ever own.

As well as the multimeter itself, you will need two short lengths of wire that fit into the holes in the GPIO sockets. If you do not have wire on hand, a couple of small uncurled paperclips will work just fine.

Run the program 06_01_digitalWrite.js in Cloud9. This program will set pin P8_10 to 3.3V:

```
// 06_01_digitalWrite.js

var b = require('bonescript');

var pin = "P8_10";
b.pinMode(pin, b.OUTPUT);
b.digitalWrite(pin, 1);
```

We are going to test the output of pin P8_10, so set your multimeter to a DC voltage range of 0 to 20V. Your meter may be different and have a range of 0 to 10V. It does not matter, as long as the top of the range is above 5V.

Push the bare-metal end of one wire into the top-right position of the right-hand GPIO connector (with the Ethernet socket at the top). Put the second wire directly four holes beneath the first. The two sockets are labeled GND and P8_10, respectively, in Figure 6-3. If you are using uninsulated wires or paperclips, bend the wires away from each other so that they cannot touch.

Figure 6-4 shows the multimeter connected to the two wires and registering 3.3V. Note that if your multimeter came with aligator- or mini-clip leads, use

**Figure 6-4**  *Multimeter measuring GPIO pin*

them; otherwise, you will have to hold the probes to the wires, pinching each between your thumb and forefinger.

Now modify the program 06_01_digitalWrite.js as follows:

```
b.digitalWrite(pin, 1);
```

to:

```
b.digitalWrite(pin, 0);
```

Save the program and run it again. Now the multimeter should register 0V.

Setting the pin to 3.3V (also called "setting it high") requires two steps. First, you have to specify that this pin should be an output. To do this, use the following line:

```
b.pinMode(pin, b.OUTPUT);
```

Then you have to set its value to 1 or 0 using this line:

```
b.digitalWrite(pin, 1);
```

Note that a variable has been used to specify the pin name. This mean that if we change our mind about which pin we are turning on and off, we only have to change the program in one place—that place being where we set the value of **pin**.

A multimeter is useful for testing, but obviously we would normally connect something like an LED to a digital output. LEDs require a series resistor to limit the current. We cover connecting up LEDs and other types of output in the next chapter.

## Digital Inputs

Just as a digital outputs can only be on or off, digital inputs detect the voltage at a pin, and if it is higher than a certain amount they indicate that the input is high. Typically you will find components such as switches connected to a digital input, but for now we will just use a short length of wire. If you have a resistor of almost any value less than a few hundred kΩ, then this will also work and provide some immunity from wiring mistakes that might otherwise damage your board.

*CAUTION*   *When trying out the examples in this section, check and double-check that the wire is between the correct holes in the connector. Looking at Figure 6-3, you can see the different supply voltages. If you accidentally bridge any one of GND, 3.3V, 5V raw, or 5V to any other socket, you may well damage your BeagleBone Black beyond repair.*

Connect a wire between one of the GND sockets on the top row of the left-hand GPIO connector and pin P9_12. Figure 6-5 shows the connection. Be careful to count the rows correctly.

Now run the program 06_02_digitalRead.js. You should see a series of zeros appearing in the console output.

```
// 06_02_digitalRead.js

var b = require('bonescript');

var pin = "P9_12";
b.pinMode(pin, b.INPUT);

function readInput() {
    var reading = b.digitalRead(pin);
```

```
        console.log(reading);
}

setInterval(readInput, 1000);
```

Now, very carefully, disconnect the end of the wire currently in a GND socket and move that end to the 3.3V socket, as shown in Figure 6-6.

The steady stream of zeros will now change to ones.

**Figure 6-5**   *Testing digitalRead 0V*

**Figure 6-6**   *Testing digitalRead 3.3V*

## Analog Outputs

Digital outputs can only be 0V or 3.3V, which is all you need if you want to turn something on or off. But if you want to control the brightness of an LED, or the speed of a motor, you need to use an **analog output**.

Only some of the pins on the BeagleBone Black are capable of producing an analog output. These are the ones marked with "PWM" in Figure 6-3. We will come back to how PWM (pulse width modulation) works later in this section. But first let's try it out.

Open the program 06_03_analogWrite.js in Cloud9 and run it:

```
// 06_03_analogWrite.js

var b = require('bonescript');

var pin = "P8_13";
b.pinMode(pin, b.OUTPUT);
b.analogWrite(pin, 0.5);
```

We are going to use the multimeter again, but this time we are going to use a different pin (P8_13). Connect the multimeter as shown in Figure 6-7.

You will see a reading of around 1.66V. This is roughly half of 3.3V. When you look at this line of code, you will see why:

```
b.analogWrite(pin, 0.5);
```

The second parameter of **analogWrite** indicates the proportion of 3.3V. Try varying this parameter between 0.0 and 1.0 and see how the reading on the multimeter changes. Figure 6-8 shows how PWM works.

When you look at it closely, the PWM signal is not actually changing the voltage between 0V and 3.3V, but rather it is sending 2,000 pulses per second. The ratio of the amount of time that the pulse is high to the time it is low is called the *duty cycle*. With an analog output set to **0.05**, the pulse will only be

**Figure 6-7** *Testing analogOutput*

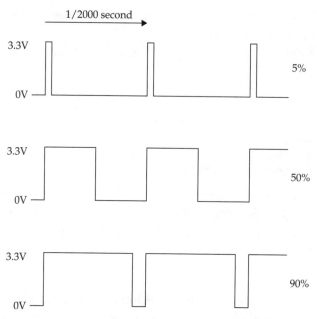

**Figure 6-8**  *Pulse width modulation*

high for a very short time. Set the duty cycle to **0.5** and the pulse will be high half the time and low half the time.

You might think that this should result in random readings of either 0V or 3.3V on the multimeter, depending on when it sampled the output. However, the multimeter does a bit of "averaging" of the signal, which is enough to even out to an approximation of an analog voltage between 0V and 3.3V.

This works just fine if we want to control the brightness of an LED. Although LEDs can react very quickly—and the LED would in fact be flashing 2,000 times a second—the human eye cannot react fast enough to see that, so the duty cycle would control the apparent brightness of the LED.

If you control something slower with PWM (say, the speed of a motor), the inertia of the rotating motor will mean that each pulse delivers a kick of power to the motor. The longer the kick (duty cycle), the faster the motor will turn.

## Analog Inputs

The analog inputs of a BeagleBone Black can read an analog voltage between 0V and 1.8V on the special analog input pins on connector P9. This is quite a narrow range, and you have to be careful not to exceed the maximum of 1.8V, or you could easily damage the board.

To test out analog inputs, all we need are two wires and an old 1.5V battery. A depleted battery is best because there is no chance of it exceeding the 1.8V limit. If you only have a fairly new one, then test it using a multimeter to make sure that it is below 1.8V before attaching it to the BeagleBone.

*CAUTION*    *Do not be tempted to use a 3V lithium cell or 9V battery—you are likely to damage your BeagleBone Black.*

Figure 6-9 shows the connections diagrammatically, and Figure 6-10 shows a photograph of the actual setup.

Open the program 06_04_analogRead.js in Cloud9 and run it. A series of readings should appear in the output console. Connect the two leads together, leaving the battery to one side to start with, and notice how the readings drop to 0. Then hold the negative lead from GND_ADC to the negative terminal of the battery and the positive lead from P9_39 (AIN0) to the positive

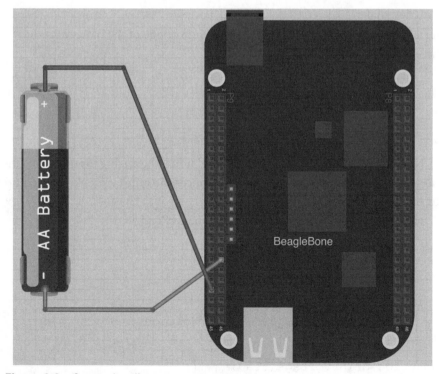

**Figure 6-9**   *Connection diagram to test analog inputs*

**Figure 6-10** *Testing analog inputs*

terminal of the battery. This will display the voltage of the battery, as shown here:

```
0
0
1.797
1.782
1.584
```

Note how the readings are roughly 1.8V when the analog input is not connected to anything. Under these circumstances the input is weakly pulled up to 1.8V.

Here is the listing for program 06_04_analogRead.js:

```
// 06_04_analogRead.js

var b = require('bonescript');

function readVoltage() {
    b.analogRead('P9_39', printVoltage);
}

function printVoltage(reading) {
```

```
    var voltage = reading.value * 1.8;
    console.log(voltage);
}

setInterval(readVoltage, 500);
```

The program defines two functions. The function **readVoltage** is responsible for reading the voltage at the analog input, and the function **printVoltage** displays the voltage in the console output. The reason why we have not just done the whole thing in a single function is that unlike **digitalRead**, **analogRead** does not just return the value that it read. Reading an analog value takes a little longer than reading a digital input, so a callback mechanism is used. Inside the **readVoltage** function, **analogRead** is called with two parameters. The first is the name of the pin to read from, and the second is the name of the function to call when the analog read has completed (in this case, **printVoltage**). The function **printVoltage** also converts the reading value from a number between 0 and 1 to an actual voltage by multiplying it by 1.8. Finally, **setInterval** is used so that the readings are made and displayed every half-second.

## Interrupts

If you look back at program 06_02_digitalRead, you will remember that we set an interval timer to repeatedly display the result of reading the pin. If we were using a digital input to detect that a button had been pressed, we could use the same approach, continually checking to see what the state of the digital input was and performing some action when it changed input value. Clearly, this is not very efficient because most of the time the computer will just be checking for a change. Interrupts provide a much more efficient way of doing this kind of thing.

You can attach an interrupt-handing function to a digital input so that if the digital input changes from high to low, the interrupt function will be run. In the meantime, your program can be off doing other things.

To test this out, we can use the same setup as we did when using **digitalRead**—that is, a header lead connected to P9_12 that you can connect to either GND or 3.3V. Start with the lead connected as per Figure 6-5 (connect P9_12 to GND) and then run the test program 06_05_interrupts.

Each time you swap the lead that connects between P9_12 and GND or 3.3V (refer to Figure 6-6), you will get a message saying that an interrupt has occurred, as shown here:

```
INTERRUPT 1
INTERRUPT 2
INTERRUPT 3
```

Here is the listing for the program:

```
// 06_05_interrupts.js

var b = require('bonescript');

var pin = "P9_12";
b.pinMode(pin, b.INPUT);
b.attachInterrupt(pin, true, b.CHANGE, interruptCallback);

var interruptCount = 1;

function interruptCallback(x) {
    console.log("INTERRUPT " + interruptCount);
    interruptCount ++;
}
```

The pin is defined as an input in the normal way and then the following line is used to attach an interrupt function to it:

```
b.attachInterrupt(pin, true, b.CHANGE, interruptCallback);
```

The first parameter is the pin to which the interrupt is going to be attached. The second parameter (**true**) indicates that the interrupt handler should always be run if there is an interrupt. This seems a little redundant. I'm sure it seemed like a good idea when the interface for interrupts was being defined. You can just set it to **true**. The third parameter specifies the type of

| Interrupt Mode | Description |
| --- | --- |
| RISING | In this case, an interrupt will only be triggered if the pin goes from low to high. |
| FALLING | In this case, an interrupt will only be triggered if the pin goes from high to low. |
| CHANGE | In this case, an interrupt will be triggered if the pin changes in either direction (that is, from high to low or from low to high). |

**Table 6-1**  *Interrupt Modes*

activity on the pin that should cause an interrupt. The choices are summarized in Table 6-1.

The final parameter to **attachInterrupt** is the function to be called when the necessary change to the pin has occurred.

The test program also keeps track of the number of interrupts detected, displaying that as part of the message to the console output when an interrupt is triggered. Generally speaking, if you have a program that needs to detect key presses, you will want it to detect them the whole time that the program is running. If this is not the case, and there are periods when you want to ignore interrupts, you can detach the interrupt from the pin using the command **detachInterrupt**, which takes the single parameter of the pin to which the interrupt was attached.

# Summary

This chapter covered the basic input/output features of analog and digital reads and writes as well as interrupts, while deliberately keeping the hardware for testing all this as simple as possible.

In the next chapter we will take these basic concepts and see how they apply to attaching real hardware components to a BeagleBone Black.

# 7

# Hardware Interfacing

In Chapter 6, we concentrated on how the BoneScript software allows you to read and write from digital and analog inputs and outputs. In this chapter, we turn our attention to hardware and how we can use LEDs, switches, and other components with the BeagleBone Black.

## Solderless Breadboard

Whereas in Chapter 6 we restricted ourselves to nothing more sophisticated than the odd piece of wire and a multimeter, in this chapter we will be attaching more interesting things to the BeagleBone Black.

A great way of doing this is to use a solderless breadboard. As the name suggests, a solderless breadboard (or just breadboard) allows you to prototype simple circuits without having to do any soldering. You simply push the components into the breadboard and then use solid-core wires or jumper wires to connect everything up and link the breadboard to the GPIO pins of the BeagleBone Black.

You can buy breadboard starter kits that provide you with a breadboard, jumper wires, and some basic components from various sources, including a kit designed specifically for this book (www.monkmakes.com). Refer to Appendix A for more information. Figure 7-1 shows how a breadboard works.

Beneath the holes on the surface of the breadboard are clips that grip any component leads or wire that are pushed through from above. Most of the holes are arranged as rows of five that are all connected together. These are arranged in two columns. The breadboard pictured in Figure 7-1 is often called a "half-breadboard." This is a useful size and big enough for anything

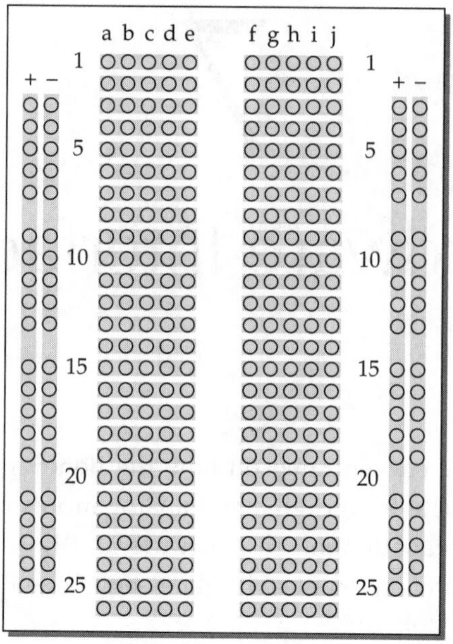

**Figure 7-1** *Breadboard construction*

we are going to build in this book. This breadboard has special columns down the left and right side of the board. These are all connected together in four single columns along the whole height of the board and are color-coded red and blue. These are often used to connect GND and +3.3V or +5V to, as a jumping off place to other connections on the breadboard. GND is connected to the negative (– blue) rail and your +3.3V or +5V power source is connected to the positive (+ red) rail.

Although solid-core wires work perfectly well in connecting one place on the breadboard to another, they are not so good for making connections between the breadboard and the BeagleBone Black because they are usually too thin to fit into the sockets on the BeagleBone Black tightly. A good alternative to solid-core wires is to use *jumpers,* which are short lengths of multi-core wire (less likely to break than solid-core wire) with connectors on each end. The type of jumper wires you need are called male-to-male jumpers because they have a little plug on each end.

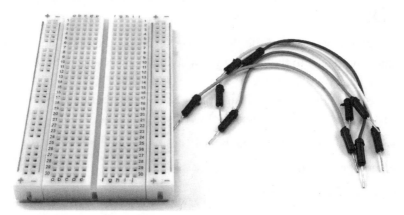

**Figure 7-2**   *Breadboard and jumper leads*

Figure 7-2 shows a half-size breadboard with a selection of jumper leads. If you are going to experiment with electronics, I strongly recommend that you obtain a breadboard and some jumper leads.

A more expensive solution is to use a breadboard cape, such as the Circuito Breadboard Cape (see Appendix A). This cape, shown in Figure 7-3, fits over the BeagleBone Black and provides an area to attach a very small self-adhesive breadboard (included). The cape also includes a couple of switches and LEDs that can be "patched" to the pins on the BeagleBone Black.

**Figure 7-3**   *A breadboard cape*

# LEDs

In Chapter 4 we used a little BoneScript to control the LEDs built onto the BeagleBone Black. In this section we will attach an external LED to the BeagleBone Black.

To use an LED with the BeagleBone Black, you will need the following items (see Appendix A for details of where to get these parts):

- Half-sized solderless breadboard
- Jumper leads (male to male)
- A red LED
- A 470Ω resistor

Connect them together as show in Figure 7-4.

The LED has one lead that is longer than the other. This is the positive lead and should be toward the top of the breadboard. The resistor is required to limit the amount of current flowing through the LED. Without such a resistor, the maximum current of the GPIO output pin would be exceeded and could damage the BeagleBone Black. The value of resistor we will use is 470Ω.

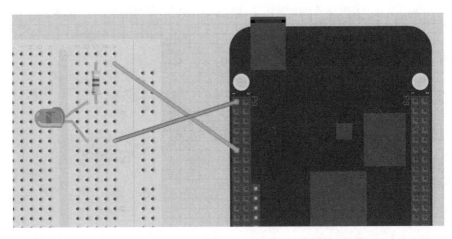

**Figure 7-4**  *Wiring diagram for a single LED*

## LEDs and Resistors

Selecting the right value of resistor for your LED requires the use of Ohm's law, which states that V = I * R. That is, the voltage across the two leads of a resistor is equal to the current flowing through the resistor (I) multiplied by the value of the resistor (R). If we know the current we need to stay below to avoid damaging the BeagleBone Black, and we know the voltage across the resistor (we can calculate this), we can rearrange Ohm's law to be R = V / I.

As you can see in Figure 7-5, LEDs have the strange property that whatever the current is flowing through them, there is an almost constant voltage across them. For a red LED, this is about 1.8V; for other color LEDs this tends to be a little higher. If we work on the basis that the LED will take 1.8 V of the 3.3V from the BeagleBone Black GPIO pin, then there will be 1.5V across the resistor (3.3 – 1.8 = 1.5V). If we aim for the maximum current allowed of 4mA (0.004 A), then the value of resistor we need is V / I = 1.5 / 0.004 = 375Ω.

That means to be safe, we must use a resistor of at least 375Ω. In fact, even with a resistor of 1KΩ, the LED will still light, but a commonly used standard value of resistor to use is 470Ω. This value of resistor will work just fine for pretty much any kind of LED, so we will stick to using it throughout.

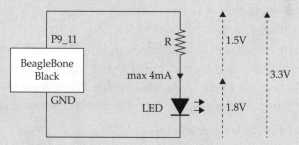

**Figure 7-5**   *Using a current-limiting resistor*

To test the LED, we could adapt any of the earlier programs we used to control the built-in LEDs. But in this case we will adapt 04_05_blink.js to use the external LED. The modified program can be found in the file 07_01_blink.js:

```
// 07_01_blink.js

var b = require('bonescript');

var led = "P9_11";
b.pinMode(led, b.OUTPUT);

var state = 0;

function toggleLED() {
    state = state ^ 1;
    b.digitalWrite(led, state);
}

setInterval(toggleLED, 500);
```

The only line of code that had to be changed is highlighted, where we changed the value of the variable **led** from **USR3** to **P9_11**.

## Switches

At its most basic, a push button switch has two contacts. When you press the button, the two contacts are connected together so that current can flow from one connection to another. A switch can easily be used with a digital input to a BeagleBone Black.

We are going to build on our earlier use of an LED to add two switches—one that increases the brightness of the LED and the other that decreases the brightness. To test out a pair of switches with a BeagleBone Black, you will need the following items (see Appendix A for details of where to get the parts):

- Half-sized solderless breadboard
- Jumper leads (male to male)
- A red LED
- Three 470Ω resistors
- Two tactile push switches

Build up the wiring for the two switches and LED as shown in Figure 7-6.

This time, we have quite a few components to connect, so we will use the supply columns down the right side of the breadboard. One is connected to GND on the BeagleBone Black, the other to 3.3V. The LED has been moved since the previous experiment with the LED. Now, it has been turned sideways so that its shorter negative lead can be connected directly to the GND supply column. The LED is also now controlled by pin P_14 because we need to use one of the pins that can perform PWM so that we can control the LED's brightness.

The switches fit across the central divide of the breadboard. The spacing of the pins is such that they will only fit the correct way across this gap because the pin spacing is rectangular rather than square. Count down the connector on the BeagleBone Black to make sure you find the right holes to put the jumper wires in. Use of a paper template such as the Bone Collar (www.monkmakes.com) makes this easier.

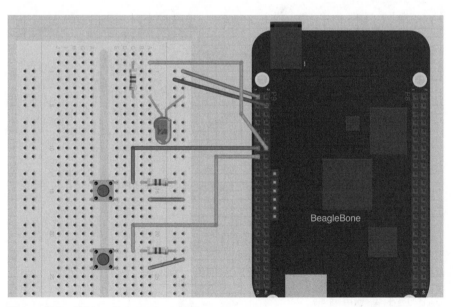

**Figure 7-6**  *Wiring diagram for switches*

## Tactile Push Switches

The type of switch we are using is called a tactile push switch (see Figure 7-7). Rather confusingly, it has four pins rather than just the two that you might be expecting. However, as you can see in Figure 7-7, the pins are connected in pairs, so we only need to connect to the two pins on the right half of the breadboard.

When the switch is not being pressed, the contacts are said to be "open." In this state, without the resistors "pulling up" the BeagleBone Black's input pins, those inputs would not be connected to anything. They would be said to be "floating." Therefore, with enough electrical interference, they could float back and forth between high and low. The resistor prevents this by pulling up the input to 3.3V. When the button is pressed, it connects the input to ground, which overrides the resistor, thus making the input low.

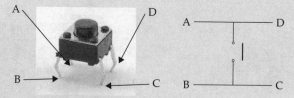

**Figure 7-7**   *A tactile push switch*

The software to control the LED brightness is listed here:

```
// 07_02_brightness

var b = require('bonescript');

var sw1 = "P9_13";
var sw2 = "P9_15";
var led = "P9_14";
var step = 0.05;

b.pinMode(sw1, b.INPUT);
b.pinMode(sw2, b.INPUT);
b.pinMode(led, b.OUTPUT);
```

```
var duty = 0.5;

function brighter() {
    duty = duty + step;
    if (duty > 1.0) {
        duty = 1.0;
    }
    b.analogWrite(led, duty);
}

function dimmer() {
    duty = duty - step;
    if (duty < 0.0) {
        duty = 0.0;
    }
    b.analogWrite(led, duty);
}

function loop() {
    if (b.digitalRead(sw1) == 0) {
        brighter();
    }
    if (b.digitalRead(sw2) == 0) {
        dimmer();
    }
    setInterval(loop, 50);
}
```

Run the program to get a feel for how the buttons control the brightness. Notice how when you hold down a button, the LED continues to get brighter or dimmer, depending on the button.

The program first defines variables for the three GPIO pins to be used. We have switched the LED over to pin P9_14 because this pin is able to produce a PWM analog output. The variable **step** is the amount by which the duty cycle of the LED is to be increased or decreased when one of the buttons is pressed.

The pins are then set to the appropriate modes of INPUT or OUTPUT and another variable called **duty** is defined. This contains the brightness of the LED as a duty cycle between 0 and 1. It is initialized to 0.5 (50 percent).

The two functions **brighter** and **dimmer** are very similar. They both change the brightness by a small amount (specified in the **step** variable) and then check to see that the value is still in the correct range of 0 to 1. Corrective action is taken if this value is too low or too high. The functions then change the brightness of the LED to the new value in **duty**.

The **loop** function is called from an interval timer every 50 milliseconds. The function **loop** checks the inputs connected to the switches and calls **brighter** or **dimmer**, as appropriate.

# RGB LEDs

RGB LEDs are colorful little devices that have three LED emitters in one plastic LED package. The LEDs are red, green, and blue. By controlling the relative brightness of these three channels, you can make the LED display pretty much any color.

To control an RGB LED, we will use three PWM-capable pins on the BeagleBone Black. For this example, you will need the following items (refer to Appendix A for details of where to get the parts):

- Half-sized solderless breadboard
- Jumper leads (male to male)
- An RGB common cathode LED
- Three 470Ω resistors

Wire up the breadboard and BeagleBone Black as shown in Figure 7-8.

When attaching the LED to the breadboard, remember that the longest lead is the common cathode, and this should be on row 2 of the breadboard.

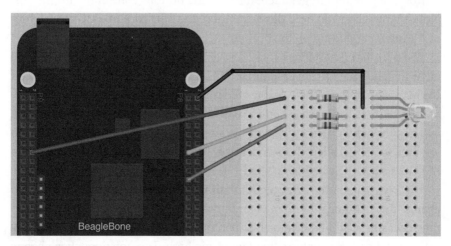

**Figure 7-8** *Connecting an RGB LED to the BeagleBone Black*

## RGB LEDs

The LED is described as "common cathode" because the negative terminals (cathodes) of the red, green, and blue LEDs are all connected together inside the LED package and are connected to the longest lead.

It is best to choose an RGBLED described as "diffuse" because it mixes the color together much better than the "clear" variety.

The following program makes the LED gradually drift from color to color:

```
// 07_03_rgb.js

var b = require('bonescript');

var ledRed = "P9_14";
var ledGreen = "P8_13";
var ledBlue = "P8_19";

b.pinMode(ledRed, b.OUTPUT);
b.pinMode(ledGreen, b.OUTPUT);
b.pinMode(ledBlue, b.OUTPUT);

var red = 0.5;
var green = 0.5;
var blue = 0.5;

function tweak(color) {
var change = (Math.random() - 0.5) / 50;
color = color + change;
    if (color> 1.0) color = 1.0;
    if (color< 0.0) color = 0.0;
    return color;
}

function changeColor() {
    red = tweak(red);
    green = tweak(green);
    blue = tweak(blue);
    b.analogWrite(ledRed, red);
    b.analogWrite(ledGreen, green);
    b.analogWrite(ledBlue, blue);
}

setInterval(changeColor, 5);
```

The colors are all started at 50-percent duty cycle. The function **change-Color** is called every 5 milliseconds, and the **tweak** function is used to change this value up or down a small amount. The **tweak** function also prevents the duty cycle value for each color from drifting outside of the range of 0 to 1.

# Switching AC

Turning devices powered from AC outlets on and off is potentially dangerous. Fortunately, a very easy-to-use module for doing just this is available, called the PowerSwitch Tail (see Figure 7-9).

The PowerSwitch Tail (see Appendix A for where to buy it) offers a safe way to control AC devices from a BeagleBone Black. The device is available for 110V U.S. outlet sockets and European 220V and 240V outlets. It has a plug on one end that fits into the electrical outlet and a socket on the other to which you connect the appliance you want to control (a desk lamp, for example).

The PowerSwitch Tail is then controlled from just two pins that are connected to the BeagleBone Black. One is ground and the other is the control signal, which is 3.3V compatible and draws around a milliamp, making it suitable for direct connection to a BeagleBone Black GPIO output. Figure 7-10 shows how you connect up a PowerSwitch Tail.

Controlling the PowerSwitch Tail is no more complicated than controlling an LED. The following program acts like a timer, turning the appliance on for a time specified in the variable **secondsOn**:

```
// 07_04_power.js

var b = require('bonescript');
var secondsOn = 20;
```

**Figure 7-9**  *The PowerSwitch Tail*

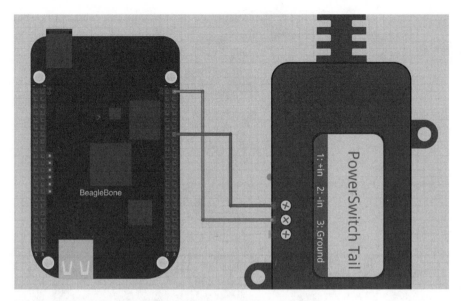

**Figure 7-10**   *Using a PowerSwitch Tail with the BeagleBone Black*

```
var controlPin = "P8_14";
b.pinMode(controlPin, b.OUTPUT);

b.digitalWrite(controlPin, 1);

function turnOff() {
    b.digitalWrite(controlPin, 0);
}

setTimeout(turnOff, secondsOn * 1000);
```

# Temperature Sensor

A practical example of using one of the BeagleBone Black's analog inputs is to attach a temperature sensor to it. The temperature sensor we will use is a three-pin device called the TMP36.

To make the thermometer, you will need the following items (see Appendix A for details of where to get the parts):

- Half-sized solderless breadboard
- Jumper leads (male to male)
- TMP36 temperature sensor

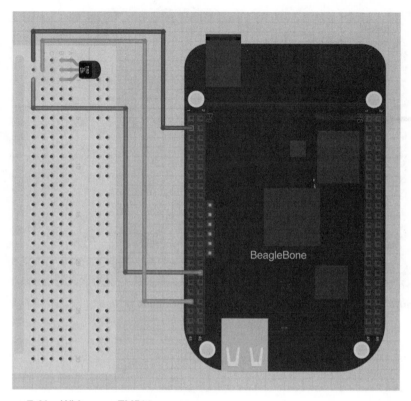

**Figure 7-11**   *Wiring up a TMP36*

Connect the temperature sensor to the breadboard, as shown in Figure 7-11. You could, if you prefer, use male-to-female jumper wires and connect the TMP36 directly to the BeagleBone Black.

The TMP36 has one flat side; make sure this is to the left, as shown in Figure 7-11. Note that rather than use the normal GND connection, we are using the GND_ADC connection. This is at the same voltage level (that is, 0V) as GND, but it's designed to be "lower noise," thus producing more accurate analog readings than if you used the normal GND connection.

Run the test program shown here:

```
// 07_05_thermometer.js

var b = require('bonescript');

function readTemp() {
    b.analogRead('P9_39', displayTemp);
}
```

```
function displayTemp(reading) {
    var millivolts = reading.value * 1800;
    var tempC = (millivolts - 500) / 10;
    var tempF = (tempC * 9/5) + 32
    console.log("Temp C=" + tempC + "\tTemp F=" +
tempF);
}

setInterval(readTemp, 500);
```

You should see a series of temperature readings in both centigrade and Fahrenheit. Try putting your finger over the temperature sensor to warm it up.

```
Temp C=20.9 Temp F=69.62
Temp C=21.1 Temp F=69.98
Temp C=21.3 Temp F=70.34
Temp C=21.4 Temp F=70.52
Temp C=21.5 Temp F=70.7
```

The program is very similar to 06_02_digitalRead.js. The function **read-Temp** is called every half-second using an interval timer. This starts off the **analogRead**, which calls back the function **displayTemp** when the analog reading is complete.

The TMP36 produces a voltage at its output that is proportional to the temperature. The temperature in degrees C can be calculated as follows:

(voltage in millivolts – 500) / 10

Fortunately, this means that unless the temperature rises above 130 degrees C (266 degrees F), the 1.8V limit will not be exceeded.

# Light Sensor

The BeagleBone Black's analog input feature can also be used in conjunction with a photoresistor to measure light intensity. To make the light meter, you will need the following items (see Appendix A for details of where to get the parts):

- Half-sized solderless breadboard
- Jumper leads (male to male)
- Photoresistor
- A 1kΩ resistor

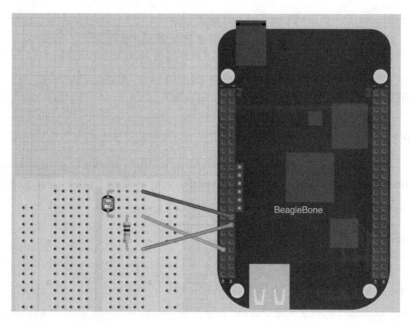

**Figure 7-12** *Wiring diagram for the light meter*

The wiring diagram for this example is shown in Figure 7-12.

It does not matter which way around you place the photoresistor. The photoresistor and the 1kΩ resistor are in an arrangement called a *voltage divider,* which uses the 1.8V ADC pin and divides it in proportion to the relative resistances of the fixed 1KΩ resistor and the photoresistor. Because the resistance of the photoresistor varies depending on the amount of light falling on it, the voltage at AIN0 pin will also vary.

Run the test program listed here:

```
// 07_06_light.js

var b = require('bonescript');

function readLight() {
    b.analogRead('P9_39', displayLight);
}
```

```
function displayLight(reading) {
    var millivolts = reading.value * 1800;
    console.log("Light=" + millivolts);
}
```

```
setInterval(readLight, 500);
```

You should see a series of readings appear. Hold your hand over the photoresistor to reduce the amount of light falling on the photoresistor and notice how the reading changes.

```
Light=124
Light=336
Light=142
Light=91
Light=92
Light=377
Light=377
```

The program is actually even simpler than the one for the temperature sensor in the previous section. The voltage at the analog input is displayed in millivolts as an indication of the brightness.

# Servos

Servo motors differ from most motors in that they are not normally designed to turn continuously. In fact, most can only turn through about 180 degrees. They incorporate feedback control and gears that allow you to set the position of the servo's rotor to a particular angle by controlling the duration of pulses on a control pin. Figure 7-13 shows how the pulse durations control the position of the servo.

A short pulse of around 1 millisecond will put the servo's rotor at one end of its travel. A 2-millisecond pulse at the other end and pulse lengths in between will put the rotor somewhere in the middle.

You are going to use a type of variable resistor called a "trimpot." When you turn the knob on the trimpot, the servo will rotate to follow the position of the knob. For this example, you will need the following items (see Appendix A for details of where to get the parts):

- Half-sized solderless breadboard
- Jumper leads (male to male)

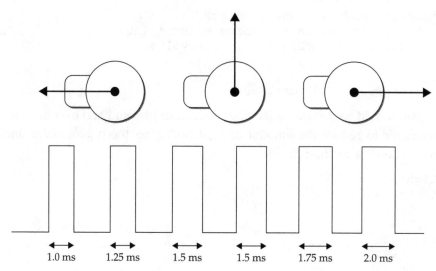

**Figure 7-13**  *How servo motors work*

- Mini servo motor (9g)

- A 10kΩ trimpot

Build the arrangement shown in Figure 7-14. Note that if you have male-to-female jumper leads, you can dispense with the breadboard entirely and connect the trimpot directly to the BeagleBone Black. Be sure to connect an external 5V power supply to your BeagleBone Black before hooking the servo up to keep it from crashing as the servo draws more current than a USB port normally provides.

Run the following program and then try turning the knob. You will see the servo motor's rotor turning in time with the knob.

```
// 07_07_servo.js

var b = require('bonescript');
var servoPin = "P9_14";
var potPin = "P9_33";
var minDuty = 0.03;

b.pinMode(servoPin, b.OUTPUT);

function loop() {
    b.analogRead(potPin, setAngle);
}
```

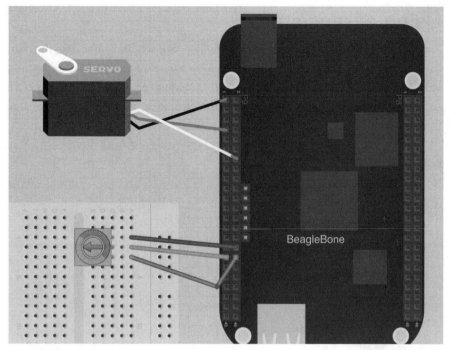

**Figure 7-14**   *Wiring up a servo and trimpot*

```
function setAngle(reading) {
var angle = reading.value * 180.0;
    var dutyCycle = (angle / 1565.0) + minDuty;
    b.analogWrite(servoPin, dutyCycle, 50);
}
```

```
setInterval(loop, 50);
```

After defining variables for the pins to be used and another value for the minimum duty cycle, we set **servoPin** to be an output.

The main **loop** function is run every 50 milliseconds by an interval timer. The **loop** function initiates an **analogRead** with a callback function of **setAngle**. The **setAngle** function first calculates the angle (0 to 180) by multiplying the analog reading (0 to 1) by 180. It then calculates the duty cycle that will generate pulses of just the right length for that angle. Finally, **analogWrite** is used to set the PWM output to that duty cycle. The third parameter to **analogWrite** is the frequency of the pulses. Because the servo expects to receive a pulse every 20 milliseconds, this is set to 50 Hz (50 times per second).

## Using a Transistor

You can use a servo motor directly with a BeagleBone Black because the control pin does not actually power the motor; it just sends control signals at a very low current (probably less than 1 mA). However, if we want to power a small 6V DC motor, this is likely to require 100 mA or more. This is far greater than the 4mA that a digital output can provide.

To control the power to DC motors and other high-power devices, such as high-power LEDs, electromechanical relays, and so on, you can use a transistor. For this example, you will need the following items (see Appendix A for details of where to get the parts):

- Half-sized solderless breadboard
- Jumper leads (male to male)
- 2N3904 transistor
- A 10kΩ trimpot
- 1N4001 diode
- 1kΩ resistor
- Small 6V DC motor

Build up the breadboard as shown in Figure 7-15.

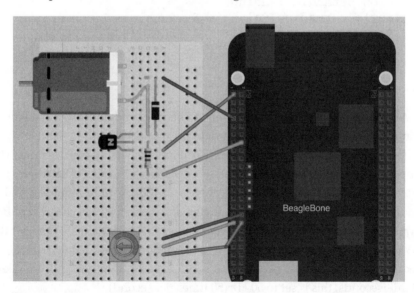

**Figure 7-15** *Breadboard layout for using a transistor*

When connecting the diode, notice that it has a stripe on one end. Make sure you connect the diode with the stripe toward the top of the breadboard. The transistor has a flat side, and this should be facing toward the right of the breadboard.

Note that driving a DC motor directly from a BeagleBone Black will only work if you are using a small motor. Larger motors are likely to draw too much current and cause the BeagleBone Black's supply voltage to dip, causing a reset. Larger motors may also burn out the transistor or damage the BeagleBone Black, so select a small motor with similar specifications to the one recommended in Appendix A.

The following program uses the trimpot to control the speed of the motor:

```
// 07_08_transistor.js

var b = require('bonescript');
var outputPin = "P9_14";
var potPin = "P9_33";

b.pinMode(outputPin, b.OUTPUT);

function loop() {
    b.analogRead(potPin, setSpeed);
}

function setSpeed(reading) {
    var dutyCycle = reading.value;
    b.analogWrite(outputPin, dutyCycle, 50);
}

setInterval(loop, 50);
```

The code is very similar to the servo program (07_07_servo.js) but is actually simpler, with the **setSpeed** function simply passing on the analog reading (0 to 1) as the duty cycle of the PWM output.

Although you have used a motor here, the transistor can be used to control other devices such as high-power LEDs, electromechanical relays, and so on. Just swap out the motor for the other device. If the device is an "inductive load" (that is, it is a motor or anything with a coil in it like a relay), then you need the diode, otherwise you don't. However, leaving the diode in place will do no harm; it just adds protection from voltage spikes. The maximum current that can be controlled by this transistor is 200 mA.

## Summary

In this chapter you learned about an array of techniques for attaching various kinds of hardware to a BeagleBone Black. You also discovered how to write simple programs to control the electronics.

In the next chapter we will look at using cape expansion boards and other ready-made modules with the BeagleBone Black.

# 8

# Using Capes and Modules

As you saw in Chapter 7, it is perfectly possible to make you own electronics on a breadboard and attach it to a BeagleBone Black using jumper wires. You can also use ready-made "capes" that plug onto the GPIO headers, or you can use other separate electronic modules that are suitable for connecting to the BeagleBone Black.

The range of capes is always expanding, so this chapter only surveys a few of the current capes. Be careful, because some of the capes available for Beagle-Bone are not compatible with the newer BeagleBone Black model. Therefore, be sure to check this before you buy. A list of capes and their compatibility with BeagleBone Black can be found at http://elinux.org/Beagleboard:BeagleBone_Capes.

## Breadboard Cape

The Breadboard Cape, shown in Figure 8-1, offers an alternative to using separate breadboard and jumper wires. You do not have to actually stick the breadboard onto the cape; you can also solder components onto the large prototyping area, in the same way as the Adafruit Proto Cape (see the next section).

You can also use the two built-in LEDs and push switches. These are broken out to a connector so that you can just bridge the connections from the LEDs and switches directly to the BeagleBone Black GPIO pins. The relationship between the header sockets and the actual components is shown in Figure 8-2.

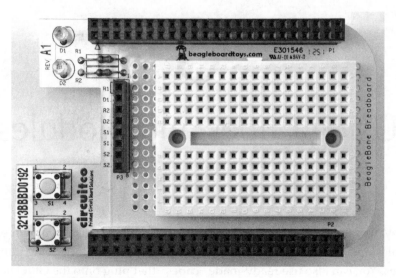

**Figure 8-1** *The Breadboard Cape*

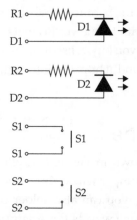

**Figure 8-2** *LEDs and switches on the Breadboard Cape*

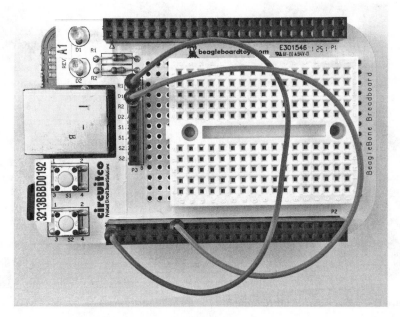

**Figure 8-3**   *Connecting up a Breadboard Cape LED*

We can wire up one of the LEDs to a GPIO pin, as shown in Figure 8-3.

The D1 (positive side of the LED) is linked to P9_14, and the R1 connection goes to GND on the BeagleBone Black GPIO connector.

The program is basically the same as 07_01_blink.js, except that it uses pin P9_14. You can find the listing in 08_01_breadboard_cape.js.

Unfortunately, the switches lack pull-up resistors, so these would need to be added to the breadboard area before they can be used.

# Proto Cape

The Adafruit Proto Cape is another prototyping board for the BeagleBone Black (see Figure 8-4). It is sold as a bare circuit board, to which pin headers must be soldered.

The board is sensibly laid out with a large central prototyping area. Also, the GPIO rows are labeled with numbers, making it easy to work out which pin is which without having to count up the rows.

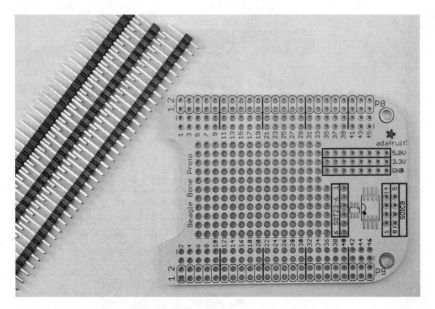

**Figure 8-4** *The Adafruit Proto Cape*

There is also an area where you can solder an SOIC 8-pin surface mount IC and an even smaller 6-pin SOT23 device. In both cases, the pins are broken out to standard 0.1-inch-spaced solder pads on the board. You might want to use such an IC for, say, an RTC (Real-Time Clock) chip.

# Battery Cape

The Battery Cape serves no purpose other than to provide the BeagleBone Black with power from four AA battery cells.

The GPIO connectors have pins on the underside of the board and sockets on the top so that the Battery Cape can become the filling in a BeagleBone Black sandwich when another cape is put on top of it.

# LCD Cape

Various sizes of LCD Capes are available for the BeagleBone Black. They provide an LCD touch screen that acts just like a monitor attached to the Beagle-Bone Black. The various models are summarized in Table 8-1.

| Name | Screen Size (Diagonal) | Resolution |
| --- | --- | --- |
| LCD3 | 3.5 inches | 320×240 |
| LCD4 | 4.3 inches | 480×272 |
| LCD7 | 7 inches | 800×480 |

**Table 8-1**  *LCD Capes*

Figure 8-5 shows an LCD3 Cape.

The LCD Cape functions just like a monitor, and the touch screen behaves like a mouse. The only problem with this is that the screen resolution is low, and the touch screen requires calibration.

The first time you boot your BeagleBone Black with the cape attached, it will perform a touch-screen calibration. A cross is displayed that you click on, and then the screen moves to the next location. You should do this with a stylus for accuracy.

**Figure 8-5**  *An LCD3 Cape*

## Motor Driver Module

When it comes to controlling the direction of DC motors, it is a good idea to use a ready-made module. An example of such a module is the SparkFun motor driver (see Appendix A). This module, shown in Figure 8-6, can be used directly with the BeagleBone Black.

Figure 8-7 shows how you can connect it up to control the speed and direction of a small DC motor. It is actually capable of controlling two motors, as you will see in Chapter 10. However, here we will use just one of its channels.

To use this module with a BeagleBone Black, you will need the following items (see Appendix A for details of where to get the parts):

- Half-sized solderless breadboard
- Jumper leads (male to male)
- SparkFun motor module
- A 10 kΩ trimpot
- 6V small DC motor

**Figure 8-6** *SparkFun motor module*

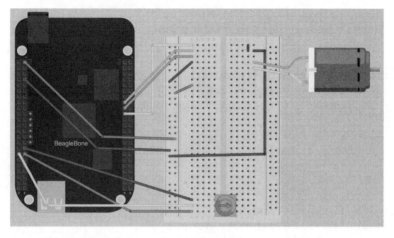

**Figure 8-7** *Wiring diagram for using a motor module*

The module needs three output pins to control each motor, and one of those pins must be capable of PWM (analog output). The PWMA pin of the motor module controls the speed of the motor. The two pins AIN1 and AIN2 control the direction of the motor. If AIN1 is high and AIN2 is low, the motor will turn in one direction, and if they are swapped over so that AIN1 is low and AIN2 is high, the motor will rotate in the opposite direction.

The test program uses the position of the trimpot to control both the speed of the motor and the direction in which it turns. At the trimpot's center position, the motor will be stopped. Turn the pot one way, and the motor will start to turn in one direction. It will rotate faster the further you turn the knob. Turn the knob the other way and the motor will reverse its direction.

```
// 08_02_motor_mod.js

var b = require('bonescript');
var pwmPin = "P8_19";
var ain1Pin = "P8_17";
var ain2Pin = "P8_15";
var potPin = "P9_33";

b.pinMode(pwmPin, b.OUTPUT);
b.pinMode(ain1Pin, b.OUTPUT);
b.pinMode(ain2Pin, b.OUTPUT);
```

```
function loop() {
    b.analogRead(potPin, adjustMotor);
}

function adjustMotor(reading) {
    if (reading.value > 0.5) {
        forwards((reading.value - 0.5) * 2);
    }
    else {
        backwards((0.5 - reading.value) * 2);
    }
}

function forwards(duty) {
    b.analogWrite(pwmPin, duty);
    b.digitalWrite(ain1Pin, 1);
    b.digitalWrite(ain2Pin, 0);
}

function backwards(duty) {
    b.analogWrite(pwmPin, duty);
    b.digitalWrite(ain1Pin, 0);
    b.digitalWrite(ain2Pin, 1);
}

setInterval(loop, 50);
```

The program is driven by an interval timer that repeatedly calls **loop**. This starts an analog reading, the callback for which (**adjustMotor**) makes the necessary adjustments to the motor's speed and direction.

The **adjustMotor** function first determines on which side of the center point the trimpot is so that it can either call **forwards** or **backwards**. It also determines the PWM duty cycle and passes that to **forwards** or **backwards**.

## I2C

I2C (pronounced *I squared C*) is an interface bus standard intended for IC-to-IC communication. It is often found on sensors and displays designed to be attached to a microcontroller. Because the GPIO connector has an I2C interface, this makes it a convenient way to connect peripherals to it. The two I2C pins are SDA and SCL, which can be found on P9_20 and P9_19, respectively.

Connecting an I2C device to a BeagleBone Black is normally as simple as connecting GND to GND, SDA to SDA, SCL to SCL and a positive power supply to a peripheral of either 5V or 3.3V, depending on the module being used.

## The node-i2c Library

Communicating with an I2C device requires the sending of serial data. To make this as painless as possible, there is a Node library designed for this purpose.

### Updating BoneScript

Before installing this or any other Node libraries, it is a good idea to check that you have the latest version of BoneScript. To update it, run the following commands:

```
# ntpdate -b -s -u pool.ntp.org
# opkg update
# opkg install python-misc python-modules
# npm config set strict-ssl false
# npm install -g bonescript
```

If the final step in this list fails with a message something like "not ok code 0," there is still a bug in the installation script and you need to edit a file by entering the following command:

```
# nano /usr/lib/node_modules/npm/node_modules/node-gyp/lib/configure.js
```

Find the JavaScript function **checkPythonVersion** and put double slashes before the lines (shown next in bold) to comment out the faulty Python version check:

```
function checkPythonVersion () {
    execFile(python, ['-c', 'import platform; print(platform.
python_version());'],
functio$
      if (err) {
        return callback(err)
      }
      log.verbose('check python version',
```

*(continued)*

```
        ' `%s -c "import platform; print(platform.python_$
    var version = stdout.trim()
    if (~version.indexOf('+')) {
      log.silly('stripping "+" sign(s) from version')
      version = version.replace(/\+/g, '')
    }
//       if (semver.gte(version, '2.5.0') && semver.lt(version,
'3.0.0')) {
      getNodeDir()
//     } else {
//     failPythonVersion(version)
//     }
    })
  }
```

Then run this command again and all should be well:

```
# npm install -g bonescript
```

To install the node-i2c library itself, open a terminal session and run the following command:

```
# npm install -g i2c
```

The library provides a number of functions. You can find the documentation for the library at https://github.com/kelly/node-i2c.

The most common way to use I2C devices is to send them a series of bytes. Often the first byte is considered to be special and identifies a command to the device or sometimes a register name. The subsequent bytes in the message contain the data. You may then request to read bytes back from the device.

Every device is different, and if you cannot find the example code for a device, you will have to refer to its datasheet to figure out the format of the messages it is expecting.

## Interfacing with an I2C Display

The first I2C example involves using an LED display module that uses the HT16K33 LED driver chip (see Figure 8-8). This chip has a load of pins for controlling an LED and just four connectors to the outside world for power and the I2C bus.

**Figure 8-8**   *Using an I2C LED display*

To try out this display, you will need the following items (see Appendix A for details of where to get the parts):

- Half-sized solderless breadboard

- Jumper leads (male to male)

- Adafruit I2C 4-digit 7-segment display module

The display is connected as shown in Figure 8-9.

Once the display is attached, you can check that it is connected properly to the I2C bus using the i2cdetect utility:

```
# i2cdetect -y -r 1
     0  1  2  3  4  5  6  7  8  9  a  b  c  d  e  f
00:          -- -- -- -- -- -- -- -- -- -- -- --
10: -- -- -- -- -- -- -- -- -- -- -- -- -- -- -- --
20: -- -- -- -- -- -- -- -- -- -- -- -- -- -- -- --
30: -- -- -- -- -- -- -- -- -- -- -- -- -- -- -- --
40: -- -- -- -- -- -- -- -- -- -- -- -- -- -- -- --
50: -- -- -- -- UU UU UU UU -- -- -- -- -- -- -- --
60: -- -- -- -- -- -- -- -- -- -- -- -- -- -- -- --
70: 70 -- -- -- -- -- -- --
root@beaglebone:~#
```

You can connect a theoretical maximum of 128 devices to the I2C bus as long as they all have a different I2C address. The address is usually fixed for

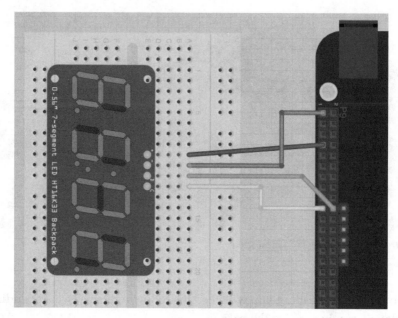

**Figure 8-9** *Wiring diagram for the I2C display module*

the device, or it can be selected from a range using jumper pads on the module. For example, the display module used has an address of 70 (in hexadecimal).

The code to use this display has been split out into a library. The library is a port of a Python library developed by Adafruit to accompany their module. It has most of the same function names, but is written in JavaScript rather than Python. The file is named ht16k33.js. Open this in Cloud9 while we examine the code, starting with the library file.

The variable **address** is assigned the value $0 \times 70$ (hexadecimal) to agree with the I2C address of the module.

The display uses the concept of registers, which control the operation of the display. Each I2C device is different, and you will normally have to track down the datasheet for the chip that the module uses. In this case, you can find the datasheet for the HT16K33 here at http://www.adafruit.com/datasheets/ ht16K33v110.pdf.

Each of these registers has an identifying byte, and variables are used to reference these registers. The registers are **dispReg** (display register), **sysReg** (system register), and **dimReg** (dimming register).

The next line sets up I2C communication with the device at the address specified:

```
var wire = new i2c(address, {device: '/dev/i2c-1', debug: false});
```

We then have a buffer of 16 bytes, each bit of which can control an individual LED:

```
var buffer = [0,0,0,0,0,0,0,0,0,0,0,0,0,0,0,0];
```

The **digits** array contains the seven-segment patterns for each of the digits, 0 to 9.

The function **start** interacts with several registers to actually start up the display itself. It first writes the byte 0 to the system register OR-ed with 1 (which amounts to adding 1 to it). It then calls **setBlinkRate** and **setBrightness**, which write to the "display" and "dim" registers, respectively.

The LED segments of the display are laid out in a grid electrically, even if they are not physically in a grid. Each digit of the display is a row of the LEDs in electrical terms. The **setBufferRow** function allows the segments of one display digit to be altered separately from the others. Its parameter is a 16-bit number, eight bits of which control segments of one digit. It manipulates the buffer appropriately, splitting the 16 bits between two bytes of the buffer and then calls **writeDisplay**, which simply sends an initial byte of 0 followed by all 16 bytes of the buffer.

The function **writeDigitRaw** is just a synonym for **setBufferRow** and is provided so that you can control the LED segments to make different characters than the stock numbers 0 to 9 and A to F (for hexadecimal).

The function **writeDigit** allows you to specify the digit position and a number to display there. Note that the colon LEDs are digit 2, so the actual numeric digits are at positions 0, 1, 3, and 4.

```
exports.writeDigit = function(charNumber, value, dot) {
    setBufferRow(charNumber, digits[value] | (dot << 7));
}
```

The **value** parameter should be between 0 and 15, where 10 will be displayed as A, 11 as B, and so on. If the **dot** parameter is true, then the dot segment of that digit will be illuminated along with the number specified.

To control the colon in the center of the display, use **setColon** to interact with digit 2.

Because nearly all the work is happening in the library, the actual clock program itself is very short:

```
// 08_03_i2c_clock.js

var d = require('./hdl6k33');

d.start();

var colonState = true;

function displayTime() {
    var date = new Date();
    var h = date.getHours();
    var m = date.getMinutes();
    d.writeDigit(0, Math.floor(h / 10));
    d.writeDigit(1, h % 10);
    d.writeDigit(3, Math.floor(m / 10));
    d.writeDigit(4, m % 10);
    d.writeDisplay();
}

function flashColon() {
    d.setColon(colonState);
    d.writeDisplay();
    colonState = ! colonState;
}

setInterval(displayTime, 1000);
setInterval(flashColon, 500);
```

Two interval timers are used. One updates the time, refreshing the display (**displayTime**) and the other (**flashColon**), toggles the colon from lit to unlit every half second.

The only unusual code is in **displayTime**, where the separate digits of the time are separated from the hours and minutes. The % operator gives the module remainder of the value to its left, when divided by the value to its right. For example, if the hour was 12, then the remainder of 12/10 is 2, which is the second hour digit. The **Math.floor** function is necessary to prevent the first digit of the hour or minute becoming a floating point number by rounding the number down to the nearest whole number.

## Interfacing with Serial GPS

Another mechanism commonly used to connect peripherals is the serial interface. Unlike a "bus" mechanism like I2C, you can only use it to connect two devices together, not to chain together a whole load of devices. The BeagleBone Black has five serial ports available for use. Each serial port has an RX (receive) and TX (transmit) pin. These operate at 3.3V, so be careful to check that the device you connect also operates at this voltage and not 5V.

Serial interfaces are more common in peripherals that might be connected directly to a computer such as GPS (Global Positioning System) and RFID (Radio Frequency ID) modules. This harkens back to a time when PCs had a serial port rather than using just USB. However, the standard is still popular.

To try out this GPS module, you will need the following items (see Appendix A for details of where to get the parts):

- Half-sized solderless breadboard
- Jumper leads (male to male)
- Adafruit Ultimate GPS Breakout

Figure 8-10 shows a GPS module wired up to the BeagleBone Black, and Figure 8-11 shows the wiring diagram.

**Figure 8-10**   *GPS and BeagleBone Black*

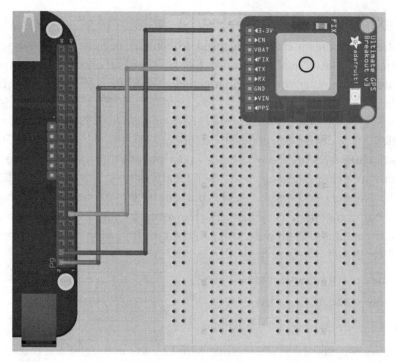

**Figure 8-11**  *Wiring a GPS module to a BeagleBone Black*

In this case, the communication is one way, from the GPS module to the BeagleBone Black. So, only the TX pin of the GPS module is connected to the RX pin of the BeagleBone Black.

Serial communication is contained in a library that you can load with the following command (see the previous section on updating to the latest version of BoneScript):

```
# npm install -g serialport
```

The test program for this GPS hardware can be found in the file 08_04_serial_gps.js:

```
// 08_04_serial_gps.js

var sp = require("serialport");

var port = '/dev/ttyO4';
var options = { baudrate: 9600, parser: sp.parsers.readline("\n") };
```

```
var gpsPort = new sp.SerialPort(port, options);

gpsPort.on('data', function(data) {
    //console.log('data received: ' + data);
    var parts = data.split(",");
    if (parts[0] == "$GPRMC") {
        var lat = parseFloat(parts[3]) / 100;
        var ns = parts[4];
        var lon = parseFloat(parts[5]) / 100;
        var ew = parts[6];
        console.log("Lat=" + lat + ns + " long=" + lon + ew);
    }
});
```

When you run the program, you will see output similar to this:

```
Lat=NaN long=NaN
Lat=53.426045N long=2.397911W
Lat=53.42604N long=2.397901W
```

The GPS module will need a good view of the sky. You may need to put it out of a window. Until it acquires a satellite, the latitude and longitude will be reported as NaN (Not a Number).

The **options** variable contains two values. The **baudrate** is the speed at which the communication occurs, and this must match the speed of the GPS device. The **parser** option specifies the way the data is read. We have chosen the option to read the data a line at a time.

The communication link is created by the following line:

```
var gpsPort = new sp.SerialPort(port, options);
```

The remainder of the code defines a handler function to be run whenever a new line of data arrives from the GPS module. The module sends an update every second. That data is sent as a number of lines in different formats. The whole set is transmitted every second. The format of the data looks like this:

```
$GPRMC,092610.000,A,5342.6381,N,00240.0134,W,8.44,277.28,161213,,,A*7A
$GPVTG,277.28,T,,M,8.44,N,15.64,K,A*0B
$GPGGA,092611.000,5342.6380,N,00240.0155,W,1,5,4.62,-
38.2,M,49.0,M,,*55
$GPGSA,A,3,06,22,18,27,03,,,,,,,,4.72,4.62,0.97*04
```

Each line starts with a **$** and a message identifier. The one we are interested in is **$GPRMC**. The fields of the data are separated by commas, and you can see (highlighted in bold) the latitude and longitude in 1/100ths of a degree.

The handler function splits the row of data into an array of strings. The fields contain the latitude and longitude as well as the North/South and East/West indicators.

## Summary

In this chapter, you learned some techniques for using BeagleBone capes as well as I2C and serial modules. It is not possible to cover all the possible types of devices that could be attached to a BeagleBone Black.

In the next chapter, we will look at creating web-based user interfaces to work with the hardware attached to the BeagleBone Black.

# 9

# Web Interfaces

**Controlling electronics** through BoneScript is fun, but imagine just how much more fun it would be if we could use a web page as our user interface. This would allow us to link buttons, controls, and text on a web page to electronics attached to the BeagleBone Black. What's more, we could connect to that web page using a mobile device.

In this chapter, we look at how to do the following:

- Control digital outputs using web page hyperlinks
- Set analog outputs using a slider
- Continuously display analog and digital inputs on a web page

By having the BeagleBone Black serve a web interface to allow the control of electronics, you can access the hardware over your local network and even over the Internet with suitable modifications to you home router settings to open up port forwarding.

## How Does It Work?

Although not much code needs to be written to make a web interface, the way it works can take a little getting used to. Your little BeagleBone Black is going to be hosting a web server as well as maintaining a separate communication link with whatever computer is running the browser. Figure 9-1 summarizes this.

Starting with the web server side of things, the BeagleBone Black will run a few lines of Node JavaScript that allow it to behave as a web server. That is, it will listen, waiting for a request to arrive. A request will arrive (step 1 in

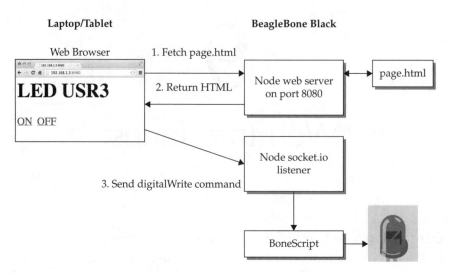

**Figure 9-1**   *Web interfaces to the BeagleBone Black*

Figure 9-1) when you type the URL of the BeagleBone Black into your browser window using the port 8080.

When this happens, the web server code on the BeagleBone Black reads the contents of the file (in this example, represented by the box "page.html"). The contents will be streamed back to the browser, which will display them (step 2).

As part of the page-loading process, some code on the page just fetched by the browser will establish a two-way communication link with a second server component on the BeagleBone Black. This allows click events on the page to send messages directly to the BeagleBone Black without having to reload the page. This is called a "socket connection" and uses the Node library socket.io.

Nothing further will happen until you click the ON or OFF link. When you do, a message will be sent over the socket connection and handled by the program running on the BeagleBone Black. This program will use the relevant BoneScript command to turn the LED on or off.

It sounds like madness, but the BeagleBone Black holds a web page containing code that the browser fetches and displays. Then, when ON or OFF is clicked, it runs that code on the browser, which sends a message to some corresponding code running on the BeagleBone Black that then turns an LED on or off.

If you are not used to web programming, this will probably sound somewhat confusing. Therefore, after introducing some basic HTML, we will follow up with a full example of controlling an LED with a web interface.

Although we will examine each type of interface (digital and analog inputs and outputs) in isolation, these can easily be combined to make complex user interfaces, which you will also learn how to do.

## Some Basic HTML

HTML (HyperText Markup Language) is the language used to create web pages. You can view the HTML of most web pages in your browser by right-clicking the browser window and selecting View Source. If you do this, you will see that the language is made up of a mixture of structural "tags" enclosed in angle brackets (< and >) as well as the actual text that appears on the page.

HTML elements have a start tag and an end tag. The start tag looks like this:

```
<tag>
```

The end tag looks like this:

```
</tag>
```

Note the added slash (/) before the name of the tag.

Every web page should be enclosed entirely within an "html" tag and contain a "head" and "body" tag, like this:

```
<html>
   <head>
   </head>
   <body>
   </body>
</html>
```

Tabs are often used to make the HTML easier to read. The "head" tag is concerned with JavaScript and information about the page that is not directly displayed, whereas the "body" tag should contain items related to what will be displayed on the page.

**Figure 9-2** *Hello World in a browser*

The following is perhaps the simplest example of a web page. You can, if you like, save this to a text file with the extension ".html" and then double-click it to open it in a browser. This works fine, as browsers are just as happy displaying pages in files on your computer as files on some remote web server. It will look something like what's shown in Figure 9-2.

```
<html>
  <head>
  </head>
  <body>
    <h1>Hello World</h1>
  </body>
</html>
```

We have added a new tag (**<h1>**) in the body of the web page. This tag indicates a level-one heading. This formatting command tells the browser that it should display the text contained in the tag in a large, bold font. It says nothing about the actual fonts or font sizes. The original concept behind HTML and the World Wide Web was that this should be enough information for the browser. However, this thinking did not last long, and fine control of a page's appearance became a requirement of HTML. In a later section you will see how to gain greater control over the page's appearance.

As you go through the following sections, you will discover some more tags that we will use to create the web interface for our electronics.

## On/Off LED Control

The first example we will look at involves turning the built-in LED USR3 on and off using the web browser interface shown in Figure 9-3.

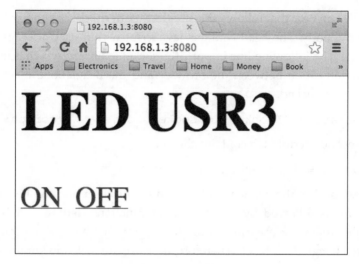

**Figure 9-3**   *A web interface to control an LED*

Two files are needed to make this interface, and you can find them among the code examples for the book in the files 09_01_led_control.html and 09_01_led_control_server.js.

Let's start with the HTML file that the BeagleBone Black will serve to the browser, which will then display it:

```
<html>
<head>
    <script src = "/socket.io/socket.io.js" > </script>
    <script>
    var socket = io.connect();
    function ledOn(){
        socket.emit('digitalWrite', '{"pin":"USR3", "value":1}');
    }
    function ledOff(){
        socket.emit('digitalWrite', '{"pin":"USR3", "value":0}');
    }
    </script>
</head>

<body>
    <h1 style="font-size:60pt;">LED USR3</h1>
    <a href="" onClick="ledOn();" style="font-size:32pt;">ON</a>
       
    <a href="" onClick="ledOff();" style="font-size:32pt;">OFF</a>
</body>
</html>
```

As you can see, this follows the standard pattern, described earlier, of a head and a body. The head contains JavaScript within **<script>** tags in the HTML file. The following line imports the socket.io library, which is used for the communication between the browser and the BeagleBone Black when ON or OFF is clicked.

```
<script src = "/socket.io/socket.io.js" > </script>
```

The second "script" tag contains the line

```
var socket = io.connect();
```

which establishes the connection between the browser and the BeagleBone Black. This is followed by two JavaScript functions (**ledOn** and **ledOff**). Remember that these are running on the browser on your normal computer, not on the BeagleBone Black. Therefore, they use the socket.io library to send a message to the BeagleBone Black.

```
socket.emit('digitalWrite', '{"pin":"USR3", "value":1}');
```

The message is sent using the **emit** function call in socket.io. This takes two parameters: the first is the command (in this case, **digitalWrite**), and the second is a parameter for the command being sent. This text may look familiar if you think back to the objects in Chapter 5. It is in fact the string representation of an object that has two values: **pin** and **value**. The "pin" is the GPIO pin (or in this case, the built-in LED), and the "value" is 1 or 0 to turn that pin on or off.

The body section of the HTML is concerned with the visual contents of the page. The first line displays an "h1" header. However, we have added extra "styling" commands to make the font 60 point:

```
<h1 style="font-size:60pt;">LED USR3</h1>
```

Next, we find the HTML for the ON hyperlink:

```
<a href="" onClick="ledOn();" style="font-size:32pt;">ON</a>
```

This uses an "a" tag. The attribute of this tag, called "href," makes the text enclosed in the tag act like a hyperlink, so we can see it change in appearance when we click it. You could use an image or many other types of HTML tags here. The important thing is that the tag contains the **onClick** attribute. This names a function to be called when ON is clicked. Not unsurprisingly, this is the **ledOn** function you met earlier.

Remembering that all this happens on the browsing computer, not the BeagleBone Black, let's turn our attention to the code running on the Beagle-Bone Black, which is listed next, broken into sections. You may find it useful to have this code open in Cloud9 while we walk through it.

The code starts by importing the modules it requires. The "http" and "fs" (file system) libraries are needed for the web server part of the code, socket.io is needed for the socket communication with the browser, and you already know about the bonescript module.

```
// 09_01_led_control_server.js

var app = require('http').createServer(handler);
var io = require('socket.io').listen(app);
var fs = require('fs');
var bb = require('bonescript');
```

The program now defines a variable for the web page that is to be served when a web request comes in and then starts the web server listening for requests on port 8080. Note that port 8080 is used to avoid a conflict with port 80 (normal web serving) and port 3000 (Cloud9). This is the reason for the ":8080" on the end of the URL when you use the web interface for controlling the LED.

```
var htmlPage = 'Prog BBB/09_01_led_control.html';
app.listen(8080);
```

The **handler** function (in the next section of code) is responsible for handling any incoming web requests. When it receives one, it reads the contents of the file specified in the variable **htmlPage** and sends it to the browser. This is not entirely obvious from the code, but here is how it works: In typical JavaScript style, the call to **readFile** specifies a callback for the read being completed. This callback function is supplied as the second parameter to **readFile**. Assuming that there is no error (usually caused by the HTML file being absent), the **res** (response) parameter to **handler** is used to tell the browser that the page is okay (status code 200). Then **end** is used to end the request and at the same time send it the whole contents of the file.

```
function handler(req, res) {
  fs.readFile(htmlPage,
    function (err, data) {
      if (err) {
        res.writeHead(500);
```

```
    return res.end('Error loading file: ' + htmlPage);
  }
  res.writeHead(200);
  res.end(data);
});
}
```

The remainder of the server code is responsible for handling the socket-based communications—in this case, messages arriving when either ON or OFF is clicked on the web interface. The function **onConnect** will automatically be called when the corresponding browser code creates the connection. It is, in fact, the very last line of the program that ensures this happens.

Inside it, you specify the commands you are prepared to handle and associate them with a function. In this case, the only command we are interested in is **digitalWrite**:

```
function onConnect(socket) {
    socket.on('digitalWrite', handleDigitalWrite);
}
```

When this command arrives, the function **handleDigitalWrite** is called. This takes the string passed in the message parameter and turns it into an object using the function **JSON.parse**.

The **pin** and **value** parts of this object are then used with **digitalWrite** to actually turn the specified pin (or LED) on or off:

```
function handleDigitalWrite(message) {
    var data = JSON.parse(message);
    bb.pinMode(data.pin, bb.OUTPUT);
    bb.digitalWrite(data.pin, data.value);
}
```

This final line ensures that **onConnect** gets called when communication is established:

```
io.sockets.on('connection', onConnect);
```

Before you can run this example, you need to install the socket.io module. Open an SSH session or Terminal window and issue the following commands:

```
# ntpdate -b -s -u pool.ntp.org
# npm update
# npm config set strict-ssl false
# npm install -g socket.io
```

You will need to have a network connection to do this. Having a network connection will allow you to control the LED from any device with a browser that is connected to your network.

Start the server program (09_01_led_control_server.js) in Cloud9 and make sure it doesn't crash; then open a browser on your main computer using the IP address of your BeagleBone Black (for example, 192.168.1.3) followed by ":8080". For my BeagleBone Black, the full URL is http://192.168.1.3:8080.

Once you can control the built-in LED (USR3), it is an easy change to control an external LED on any GPIO pin. To use an external LED in pin P9_14 as you did in Chapter 7 (see Figure 9-4), simply wire up the LED and then change the pin in the 09_01_led_control.html file. Change the two references to "USR3" to "P9_14," as shown here:

```
function ledOn(){
    socket.emit('digitalWrite', '{"pin":"P9_14", "value":1}');
}
function ledOff(){
    socket.emit('digitalWrite', '{"pin":"P9_14", "value":0}');
}
```

To use an LED with a BeagleBone Black, you will need the following items (see Appendix A for details of where to get the parts):

- Half-sized solderless breadboard
- Jumper leads (male to male)
- A red LED
- A 470 Ω resistor

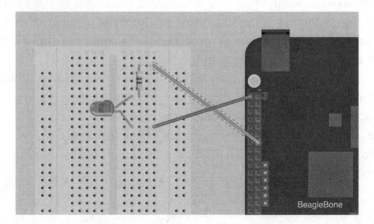

**Figure 9-4**  *Wiring diagram for an external LED*

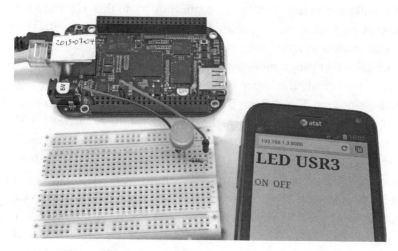

**Figure 9-5** *Android phone controlling an LED*

Note that you will have to reload the page for the change to take effect. You can now control the LED from anywhere on your network, using almost any device with a browser.

Figure 9-5 shows an Android phone connected to this address using Wi-Fi.

Note that devices with old browsers may not work, but most modern devices will be fine.

# LED Brightness Control (Analog Outputs)

You can take the same LED on a breadboard connected to pin P9_14 and control the brightness of the LED using **analogWrite**. Figure 9-6 shows the same smartphone from the previous section being used to control the brightness of the LED using a slide control.

The code for this is very similar to that of the on/off control. You will find the code in the two files 09_02_pwm_server.js and 09_02_pwm.html. Open them up in Cloud9 and follow along as the code is explained.

The HTML file is very similar to the on/off control. One difference is that the two functions **ledOn** and **ledOff** have been replaced by the single function **updateSlider**:

```
function updateSlider(value){
    socket.emit('analogWrite',
      '{"pin":"P9_14", "value":' + value / 100 + '}');
}
```

**Figure 9-6**   *Controlling LED brightness with a slider control*

This function will be called every time there is a change in the slider position, and it simply sends the slider position (0 to 100) as an **analogWrite** command to the server. Note that the range of 0 to 100 is divided by 100 to put the position into the range 0 to 1 that the server expects.

The other difference in the HTML file is that the two hyperlinks ON and OFF have been replaced by a "range" control:

```
<input type="range" min="0" max="100" step="1" value="50"
 onchange="updateSlider(this.value)" style="width: 130px;"/>
```

This control is relatively new to HTML and may not work on some older browsers. It is also limited because it is not possible to change the size of the control. Later in this section, we look at how to use a third-party control that looks better.

The **min**, **max**, and **step** values set the range and resolution of the changes in reading as you move the slider. The **onChange** attribute specifies the JavaScript function to call every time the slider position is changed and also passes it an argument of the current position.

The server code is also similar to the on/off control. The first significant difference is to the **onConnect** function, where we now need to handle **analogWrite** commands coming from the browser:

```
function onConnect(socket) {
    socket.on('analogWrite', handleAnalogWrite);
}
```

The code for **handleAnalogWrite** is very similar to that for **digitalWrite**:

```
function handleAnalogWrite(message) {
    var data = JSON.parse(message);
    bb.analogWrite(data.pin, data.value);
}
```

You may also like to look at the example file 09_03_pwm_server_fancy.js and the accompanying 09_03_pwm_fancy.html file. These use the much prettier slider shown in Figure 9-7.

This slider comes from a JavaScript module called jQuery. This module adds a lot of advanced features to HTML and is widely used. It is also quite complex, and it is beyond the scope of this book to explain jQuery. However, you can find out much more about it at jQuery.org.

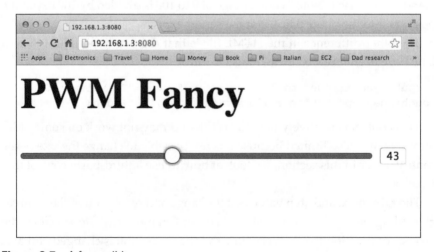

**Figure 9-7**  *A fancy slider*

## Displaying Digital Inputs

When it comes to displaying inputs, a slightly different approach is needed. Let's start with an example that uses a push switch as a digital input.

You will need the following items (see Appendix A for details of where to get the parts):

- Half-sized solderless breadboard

- Jumper leads (male to male)

- 470 Ω resistor

- Tactile push switch

Build up the breadboard as shown in Figure 9-8.

You can find the example code for this in 09_04_digital_in.html and 09_04_digital_in_server.js. Let's start with the HTML. The full listing is provided next, but it is broken up into chunks that are described separately.

```
<html>
<head>
    <script src = "/socket.io/socket.io.js" > </script>
    <script type=text/javascript
        src="http://code.jquery.com/jquery-1.7.1.min.js"></script>
```

In the header, the jQuery module is imported. In this case, this is not because we need some fancy user interface, but because it has a useful

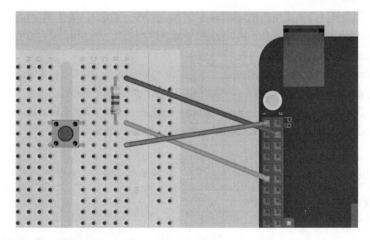

**Figure 9-8**  *Breadboard layout for a single switch*

feature that allows us to easily find a part of the HTML as it is running on the browser and manipulate this DOM (Document Object Model) so that we can change what the browser is displaying after the page has loaded. We need to do this so that we can display the status of the switch dynamically.

```
<script>
    var pin = 'P9_11';
    var socket = io.connect();
    socket.on("pinUpdate", handlePinUpdate);
    function handlePinUpdate(message) {
        var data = JSON.parse(message);
        $("#switchStatus").html("pin " + data.pin +
        " value " + data.value);
    }
    socket.emit('monitor', pin);
</script>
</head>
```

The script section of the header defines a variable for the pin to use and then opens a socket connection to the server, just like the earlier examples. It also sets up a listener for incoming messages from the browser. Any such messages will be of the type **pinUpdate** with a parameter that comprises an object containing the pin that has changed and its new value.

The syntax **$("#switchStatus").html(** uses jQuery to change the contents of the element **switchStatus** to the HTML (or just text) supplied as its parameter in parentheses. In this case, that is a string constructed from the pin name and its new value.

The function **handlePinUpdate** will be invoked each time this message is received and modifies the **switchStatus** HTML to display the state of the input, as shown in Figure 9-9.

After the function is defined, a message is sent to the server ("monitor") telling the server to report any changes to the pin specified.

The body of the page has an "h2" tag with an **id** of **switchStatus**, which is changed to display any pin changes, as described above.

```
<body>
    <h1 style="font-size:60pt;">Switch</h1>
    <h2 style="font-size:40pt;"id="switchStatus">-</h2>
</body>
</html>
```

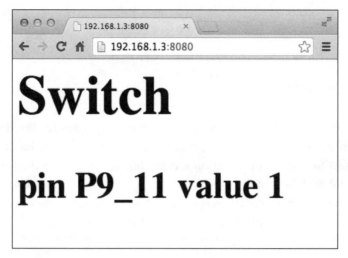

**Figure 9-9**  *Displaying the switch status*

The approach used here of informing the monitor which pin or pins the page is interested in, and then handling update messages coming back from the server, makes it very easy to monitor multiple pins. Let's now take a look at the corresponding server code.

Again, this code is broken down into sections. We will start after the usual **require** statements.

```
var pinStates = {};
var soc;
```

The first of these two variables (**pinStates**) will be used to keep track of when a pin changes. Although this is a prime candidate for the use of interrupts, at the time of writing, there are some bugs with interrupts in BoneScript. Therefore, we will have to implement an interrupt-like mechanism ourselves, and to do that we need to keep track of the previous state of a pin so that we know when it changes.

The **soc** variable provides a handle for the socket connection to the browser.

The web serving section is the same as the other examples of web interfaces, so we will skip that section.

```
function onConnect(socket) {
    socket.on('monitor', handleMonitorRequest);
```

```
    soc = socket;
}

function handleMonitorRequest(pin) {
  bb.pinMode(pin, bb.INPUT);
  pinStates[pin] = 0;
}
```

When the connection has been made, the server listens for **monitor** commands from the browser, and when one is received it runs the **handleMonitorRequest** function. This function sets the pin mode of the requested pin to be an input and then adds an entry for the pin to the **pinStates** variable.

The next section of code is responsible for checking the states of the pins and is invoked repeatedly from an interval timer:

```
function checkInputs() {
  for (var pin in pinStates) {
    var oldValue = pinStates[pin];
    var newValue = bb.digitalRead(pin);
    if (oldValue != newValue) {
      soc.emit("pinUpdate", '{"pin":"' + pin + '",
                "value":' + newValue + '}');
      pinStates[pin] = newValue;
    }
  }
}
```

```
io.sockets.on('connection', onConnect);
setInterval(checkInputs, 50);
```

The **checkInputs** function steps over each of the entries in **pinModes**. In this example, there will never be more than one entry, but the use of this data structure makes it easy to add lots of inputs. It reads the new input value and compares this with the stored value. If they differ, it sends the appropriate message to the browser, telling it to update the web page.

## Displaying Analog Inputs

Displaying analog readings is very similar to digital inputs. We could, for example, wire up a TMP36 temperature sensor, just like we did in Chapter 7 (see Figure 9-10) and display the temperature readings on a web page.

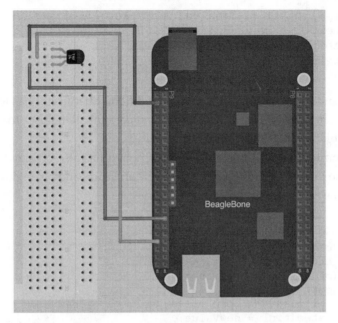

**Figure 9-10**  *Wiring diagram for a temperature sensor*

To make the thermometer, you will need the following items (see Appendix A for details of where to get the parts):

- Half-sized solderless breadboard

- Jumper leads (male to male)

- TMP36 temperature sensor (or LM35 sensor if you use the alternative programs labeled LM35 in the code download)

A good way to test out the hardware before you start on the web interface is to build the temperature sensing example from Chapter 7.

You can find the example programs files in 09_05_analog_in.html and 09_05_analog_in_server.js. When you run 09_05_analog_in_server.js and open a browser on your BeagleBone Black's IP address followed by ":8080", you will see something like Figure 9-11.

The code is almost identical to that of the previous digital input example. Therefore, we will just highlight the main differences, starting with the web page:

```
function handlePinUpdate(message) {
  var data = JSON.parse(message);
```

```
var millivolts = data.value * 1800;
var tempC = (millivolts - 500) / 10;
var tempF = (tempC * 9/5) + 32;
$("#temp").html(" " + tempF + " F");
}
```

The **handlePinUpdate** function is now different because it has to convert the analog reading between 0 and 1 into a temperature. Refer back to Chapter 7 for an explanation of the math.

Over on the server, the **checkInputs** function has also changed, because there is no longer a need to set the pin to be an input, and an analog reading needs to be taken rather than a digital one.

```
function checkInputs() {
  for (var pin in pinStates) {
    var oldValue = pinStates[pin];
    var newValue = bb.analogRead(pin);
    if (oldValue != newValue) {
    soc.emit("pinUpdate",
        '{"pin":"' + pin + '", "value":' + newValue + '}');
    pinStates[pin] = newValue;
}
```

**Figure 9-11**   *Displaying the temperature*

## Custom Actions

It makes sense for the browser to concentrate on providing the user interface, displaying data, providing buttons to click, and so on, and for the server side to have as much autonomy as possible when it comes to looking after the hardware. In the examples so far, the user interface has been directing the server to control specific pins or read values for specific pins. You may prefer to move those kinds of decisions about which pin should be used and other low-level details to the server code. For example, we could make a timer with a web interface. When we click on a button on its web page it could use a PowerSwitch tail (see Chapter 7) to turn on an appliance for ten minutes. We could arrange for all that to happen in the browser's code. However, if the browser window is closed, the pin would not turn off after 10 minutes. In this case, it's much better for the timing code to live on the server.

In this example, we have changed the delay duration times from minutes to seconds to make testing less time consuming (see Figure 9-12). The example assumes that you have wired up an LED to P9_14, as shown back in Figure 9-4.

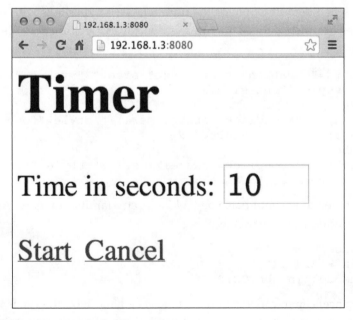

**Figure 9-12**   *A web-controlled timer*

The program files for this example can be found in 09_06_timer.html and 09_06_timer_server.js. Let's start with the HTML page. This starts just the same as the other examples in this chapter. We then come to the main script area, where we define the function **startTimer**:

```
function startTimer(){
  var duration = $("#duration").val();
  socket.emit('startTimer', duration);
}
```

This uses jQuery to find the value that is entered in the "duration" input field. It then sends the message **startTimer** to the server with a parameter of the duration.

A second function, **cancelTimer**, is used so that the message **cancelTimer** can be sent to the server when the Cancel hyperlink is clicked. This does not need any parameters.

```
function cancelTimer(){
  socket.emit('cancelTimer');
}
```

The body has an input field with an **id** of **duration**. The **value** attribute of this tag sets a default value for the duration. Two hyperlinks with **onClick** attributes are then defined to run the two JavaScript functions described previously.

```
<input id="duration" type="text" size="3" value="10"
      style="font-size:32pt;"/>
    </p>
    <a href="" onClick="startTimer();" style="font-
      size:32pt;">Start</a>
       
    <a href="" onClick="cancelTimer();" style="font-
      size:32pt;">Cancel</a>
```

The server code should also be looking pretty familiar by now. We have, however, added a few variables:

```
var cancelTimer;
var pin = "P9_14";
bb.pinMode(pin, bb.OUTPUT);
```

The **cancelTimer** variable is used to keep a handle on the timer used to cancel the pin output. This is so that if the Cancel hyperlink is clicked, the timer can be cancelled. If we did not cancel the timer but simply let it run its course, it could unexpectedly cancel a second timed period of the LED being on if that period was started right after the previous timed period was cancelled.

Because control over which pin to use is now with the server, rather than being passed as a parameter from the web page, we can set the pin's mode to output once at the top of the file.

In this case, we are looking for two possible events, so there are two lines defined in **onConnect**:

```
function onConnect(socket) {
  socket.on('startTimer', handleStartTimer);
  socket.on('cancelTimer', handleCancelTimer);
}
```

The **handleStartTimer** function turns the LED pin on and then sets a timeout that will call **handleCancelTimer** after the allotted time:

```
function handleStartTimer(duration) {
  bb.digitalWrite(pin, 1);
  cancelTimer = setTimeout(handleCancelTimer,
                           duration * 1000);
}

function handleCancelTimer() {
  bb.digitalWrite(pin, 0);
  clearTimeout(cancelTimer);
}
```

The **handleCancelTimer** can either be called by **setTimeout** or by a click of the Cancel hyperlink. It first turns off the output pin and then clears the timeout. Clearing the timer is not necessary if it was called as a result of **setTimeout**, but it will do no harm.

# Inputs and Outputs Together

All the preceding examples can be used in combination. To illustrate this, the following example uses a temperature sensor and an LED as the hardware as well as a web page that reports the temperature and also provides hyperlinks to turn the LED on and off.

To make this, you will need the following items (see Appendix A for details of where to get the parts):

- Half-sized solderless breadboard
- Jumper leads (male to male)
- TMP36 or LM35 temperature sensor
- Red LED
- 470 Ω resistor

Figure 9-13 shows the breadboard layout, and Figure 9-14 shows the web interface.

The code for this is pretty much a merge of the programs 09_01 and 09_05. You can see the completed program in the files 09_07_combo_server.js and 09_07_combo.html.

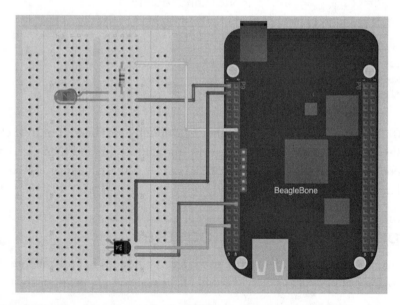

**Figure 9-13**  *Breadboard layout for the LED and temperature sensor*

**Figure 9-14**   *Web interface for the LED and temperature sensor*

## Summary

In this chapter, you learned how to build a simple web interface to interact with the BeagleBone Black. This is a very powerful technique because it allows the BeagleBone Black to be controlled from the browser of any computer on your network.

In Chapter 10, we will use this approach to make a browser-controlled roving robot.

# 10

# A Roving Robot

**Although this** is primarily a book about programming the BeagleBone Black, it can be very helpful to see how everything fits together into a real project involving hardware as well as software. This chapter and the next provide two simple projects that will help you get started building real projects with the BeagleBone Black.

The only soldering involved in this project is to attach header pins to the motor module. In the next chapter you can build a project that does require soldering of components onto the Adafruit Prototyping Cape.

The roving robot, shown in Figure 10-1, uses a Breadboard Cape, a SparkFun motor shield, and a low-cost robot chassis kit. Mobile power is provided by a 5V USB rechargeable battery unit and AA battery pack.

## Hardware

The project uses the same SparkFun motor module we used back in Chapter 8. However, this time we are using both channels to control the two motors on the robot chassis.

To make this project, you will need the following items (see Appendix A):

- Circuito Breadboard Cape
- Jumper leads (male to male)
- SparkFun motor module
- Header pins
- Magician robot chassis

- 5V 1A USB backup battery
- Barrel-jack-to-screw terminal adapter
- Wi-Fi dongle

The project uses a USB backup battery to power the BeagleBone Black. Because the project uses a USB Wi-Fi dongle, you will need a backup battery that is capable of supplying more than the usual 500mA that a computer will supply over USB. Select one that is rated at 1A or more.

## Step 1: Assemble the Magician Chassis

The first step is to assemble the magician chassis that is supplied as a kit. The kit comes with instructions. Note that other similar chassis are available.

Thread the four leads for the motors through to the top of the chassis, as shown in Figure 10-2.

## Step 2: Attach Header Pins to the Motor Module

The motor module is supplied without any connectors. You will need to solder header pins to it, as shown in Figure 10-3.

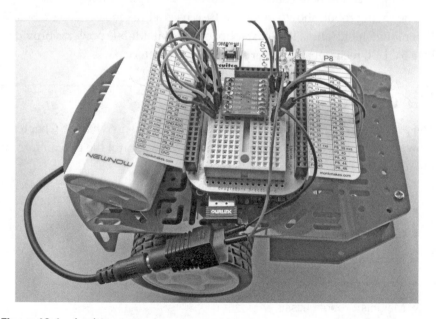

**Figure 10-1** *A robot rover*

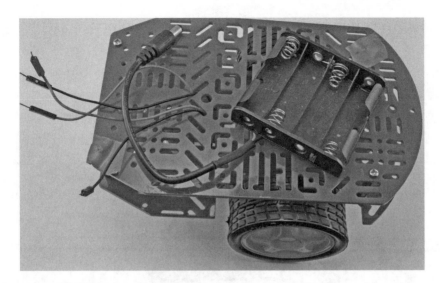

**Figure 10-2**   *The magician chassis*

**Figure 10-3**   *Soldering header pins to the motor module*

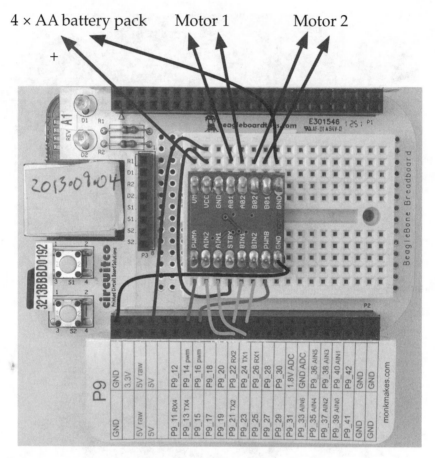

**Figure 10-4**   *Wiring the breadboard*

## Step 3: Wire Up the Breadboard

Peel off the backing paper of the mini breadboard supplied with the Breadboard Cape and attach it to the cape. Fit the motor module onto the breadboard and wire up the breadboard as shown in Figure 10-4. You do not need to attach the leads to the motors yet.

Once it is all wired up, you can attach the cape to the BeagleBone Black. Now is also a good time to attach the Wi-Fi dongle to the BeagleBone Black's USB socket.

## Step 4: Connect All the Parts

Now make the following connections to finish off the hardware side of the project:

- Connect the 6V battery pack of the magician chassis to the breadboard, as shown in Figure 10-4. You will need to use the barrel-jack-to-screw terminal adapter and attach jumper leads to the screw terminals that will then plug into the breadboard.

- Plug the motor leads into the breadboard. While testing, it is a good idea to leave the components off the chassis and turn it upside-down so that the whole thing can't drive off the edge of your desk unexpectedly.

# Software

The project uses a Wi-Fi USB dongle to communicate with the network and serve a web interface that is used to control the rover. If you have not already done so, follow the instructions in Chapter 2 on setting up Wi-Fi.

Before you use the web interface to control the robot, you can just test out the motors using the program 10_01_rover_motor_test.js. If this works okay, run 10_02_rover_server.js. Then aim a browser at the IP address of your BeagleBone Black followed by ":8080". The user interface shown in Figure 10-5 should appear. You may want to use a mobile device such as a smartphone to act as the remote control for the rover.

Open the test program in Cloud9. You will not see much here that we have not already looked at in depth in Chapter 9. Let's start by looking at the web page for the user interface (10_02_rover.html):

```
<html>
<head>
    <script src = "/socket.io/socket.io.js" > </script>
    <script>
    var socket = io.connect();
    </script>
</head>

<body>
    <h1 style="font-size:60pt;">Rover</h1>

    <table>
```

```
    <tr><td></td><td onClick="socket.emit('forwards');"
style="font-size:32pt;">^</td><td></td></tr>
    <tr><td onClick="socket.emit('left');"
       style="font-size:32pt;">&lt;</td>
         <td onClick="socket.emit('stop');"
           style="font-size:32pt;">S</td>
         <td onClick="socket.emit('right');"
           style="font-size:32pt;">&gt;</td>
    </tr>
    <tr><td></td><td onClick="socket.emit('backwards');"
style="font-size:32pt;">v</td><td></td></tr>
    </table>

</body>
</html>
```

The header JavaScript simply opens the connection, and then the **onClick** handlers attached to the four direction buttons and the "S" stop button call **socket.emit** directly.

On the server, each of these messages invokes the appropriate command to set the motor pins of the motor controller module.

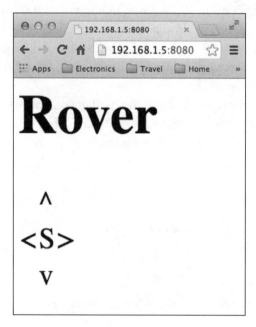

**Figure 10-5**   *A web interface for the rover*

## Summary

Apart from the need to attach pins to the motor module, this has been a largely solder-free project. In the next chapter, we will build an email notifier using a retro lamp bulb on a Prototyping Cape.

# 11

# E-mail Notifier

In this project, we combine a 12V tungsten filament lamp with the Internet capabilities of the BeagleBone Black to make an e-mail notifier that flashes the lamp when a new e-mail is received (see Figure 11-1).

The lamp used is of the type that is used for the indicator lights of a car. Although these are gradually being superseded by LEDs, these light bulbs are still easily found.

The project is built using a Prototyping Cape, and all the components are soldered into place to make the project more durable than using breadboard.

## Hardware

To build the project, you are going to need the following items. You can find more details on where to get them in Appendix A.

- Adafruit Proto Cape

- 12V 5W automotive lamp

- TIP120 transistor

- 1 kΩ resistor

- Two-way terminal block (0.2" pin spacing)

- 12V 2A DC power supply

- Barrel-jack-to-screw terminal adapter

- About six inches of solid-core insulated wire

Figure 11-2 shows the layout of the Proto Cape.

**Figure 11-1** *The e-mail notifier*

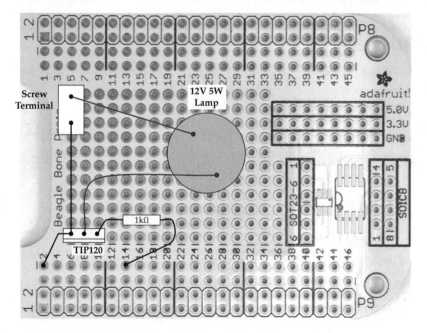

**Figure 11-2** *The Proto Cape layout for the e-mail notifier*

## Step 1: Attach Leads to the Lamp

You can obtain a lamp holder for the light bulb to make it easier to replace, or you can do what we have done here and simply solder leads to it.

The bottom connection of the light bulb will take solder easily, and you can solder a short length of solid-core wire. To make the other contact, wrap some of the solid-core wire around the bulb a few times and then solder it to the bulb body in a few places (see Figure 11-3).

## Step 2: Attach Pins to the Proto Cape

Just two connections need to be made from the Proto Cape to the BeagleBone Black. However, for stability, solder header pins to the two innermost rows of connectors (see Figure 11-4). The Proto Cape kit includes strips of header pins.

## Step 3: Solder Components to the Proto Cape

The transistor, resistor, and terminal block should now be placed onto the board and soldered to the solder pads where they emerge on the underside of the board. Make sure that the transistor is the right way around. Do not trim off the excess leads because these can be used to connect up the components under the board.

Figure 11-5 shows the topside of the board, and Figure 11-6 the underside. You can see how some of the leads have been bent to make the connections shown in Figure 11-2.

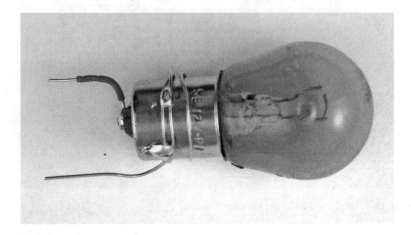

**Figure 11-3**  *Attaching leads to the bulb*

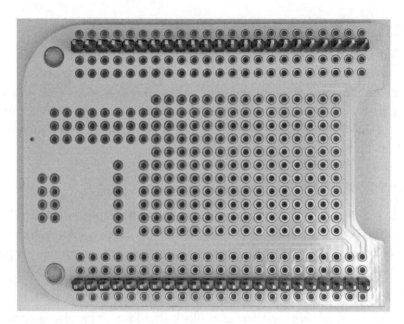

**Figure 11-4** *Soldering header pins*

**Figure 11-5** *The topside of the board*

**Figure 11-6**  *The underside of the board*

## Step 4: Attach the Lamp to the Proto Cape

Push the leads from the lamp through convenient holes in the Proto Cape and then make all the remaining connections with solid-core wire, as shown in Figure 11-7. Also, use Figure 11-2 as a reference to make sure that you have made all the connections that need to be made.

## Step 5: Connecting the Power Supply

Incandescent lamps use a lot of current—far too much for the BeagleBone Black to supply. For this reason, an external power supply is used just to supply power to the lamp. The positive side of this supply needs to be attached to the topmost screw terminal, and the negative side to the bottom screw terminal. If your power supply has trailing leads, these can just be attached to the screw terminals directly. If the power supply lead is terminated in a barrel jack plug, you can use the same barrel-jack-to-screw terminal adapter we used in Chapter 10 and then use two pieces of wire to bridge the two sets of screw terminals.

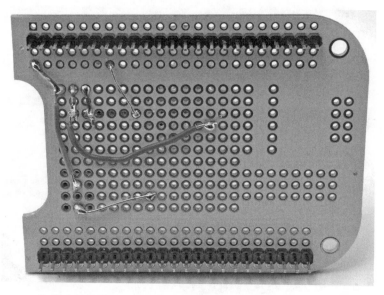

**Figure 11-7** *The completed underside of the board*

Plug the cape onto the BeagleBone Black and power it up so that we can start the software side of the project.

## Software

The project uses a library to do most of the work for the e-mail communication. To install this library, use the following command:

```
# npm install -g imap
```

Once the installation is complete, find the program 11_01_mail_notifier.js in Cloud9:

```
// 11_01_mail_notifier.js

var b = require('bonescript');
var Imap = require('imap');

var pin = "P9_14";
var maxDuty = 0.7;
var lampState = false;
var newMail = true;
```

```
var imap = new Imap({
  user: 'youremail',
  password: 'yourpassword',
  host: 'imap.gmail.com',
  port: 993,
  tls: true,
  tlsOptions: { rejectUnauthorized: false }
});

imap.once('ready', function() {
    console.log("Ready");
    imap.openBox('INBOX', true, inboxOpen);
});


function inboxOpen() {
    console.log("Inbox Open");
    imap.on('mail', notify);
}

function notify() {
    console.log("MAIL");
    newMail = true;
    setTimeout(cancelNotify, 3000);
}

function cancelNotify() {
    newMail = false;
    lampState = false;
    b.analogWrite(pin, 0);
}

function updateLamp() {
    if (newMail) {
        lampState = ! lampState;
        if (lampState) {
            b.analogWrite(pin, maxDuty);
        }
        else {
            b.analogWrite(pin, 0);
        }
    }
}

imap.connect();
```

```
setTimeout(cancelNotify, 2000);
setInterval(updateLamp, 300);
```

The variable **maxDuty** is set to control the brightness of the lamp bulb when it flashes. The variables **lampState** and **newMail** are flags indicating whether the lamp is on or off and whether new mail has been received, respectively.

In the next section, you will need to enter your e-mail details. They are currently set for Gmail, and if you want to use a different e-mail account, you will need to change **host**. Change **user** and **password** to match your account details.

The Internet Message Access Protocol (IMAP) is then initialized and a number of callback functions defined. The following line sets a callback for the receipt of a new message:

```
imap.on('mail', notify);
```

When this occurs, the **newMail** flag is set to **true**, and the timeout is set to cancel the notification process after three seconds.

An interval timer calls **updateLamp** to do any necessary flashing of the lamp, only toggling the lamp on and off as long as the **newMail** flag is **true**.

The **newMail** flag is initialized to **true** so that the lamp always flashes when the program is first run. This is just there as a check to make sure that the hardware is working. It is the purpose of the timeout on the next-to-last line of code to cancel this initial flashing of the lamp.

## Summary

In this chapter, you saw how to make use of the BeagleBone Black's network capabilities to create a fun little project. This could also be adapted for other kinds of notification—say, ringing a bell or even notifying you when someone tweets about you.

# Parts

Because this book is essentially about programming, not very many parts are referenced in it. Therefore, this chapter lists the parts that were used in the book and some possible suppliers for them.

## Suppliers

There are many suppliers of electronics and BeagleBone Black–related parts. A few of them are listed here:

| Supplier | URL | Notes |
|---|---|---|
| MonkMakes | www.monkmakes.com | Sells a kit specifically for this book. |
| Adafruit | www.adafruit.com | Their products are also stocked worldwide by local suppliers. |
| SparkFun | www.sparkfun.com | Their products are also stocked world wide by local suppliers. |
| Seeed Studio | www.seeedstudio.com | Sells unusual and low-cost modules. |
| Mouser | www.mouser.com | Offers a vast range of all types of electronic parts. |
| RadioShack | www.radioshack.com | Stocks the more common parts. |
| Digi-Key | www.digikey.com | Offers a vast range of all types of electronic parts. |
| CPC | cpc.farnell.com | UK supplier that offers a large range of parts. |
| Maplin Electronics | www.maplin.co.uk | UK supplier stocking the more common Arduino parts. |

# Breadboard Starter Kits

Breadboard starter kits are a great way of getting started with electronics. They usually comprise a breadboard, jumper wires, and some basic electronics components. They are available from many sources, including eBay and Amazon. SparkFun sell a useful kit (product ID 11038).

# BeagleBone and Capes

| Part | Suppliers (Product Code) |
|---|---|
| BeagleBone Black | Adafruit, Mouser, Amazon, Digi-Key, and others. |
| Circuito Breadboard Cape | CPC (SC12931), MCM (83-14431), Mouser (595-BB-CAPE-BREADBRD) |
| BeagleBone LCD3 Cape | CPC (SC12912), MCM (83-15521), Mouser (595-BB-BONE-LCD3-01) |
| BeagleBone LCD4 Cape | CPC (SC12919), MCM (83-15520), Mouser (595-BB-BONE-LCD4-01) |
| BeagleBone LCD7 Cape | Mouser (595-BB-CAPE-LCD) |
| Battery Cape | CPC (SC12913), MCM (83-14430), Mouser (595-BB-BONE-BATT-01) |
| Adafruit Proto Cape | Adafruit (572) |

# Modules

| Part | Suppliers (Product Code) |
|---|---|
| PowerSwitch Tail | Adafruit (268), SparkFun (10747), Amazon.com |
| SparkFun Motor Module | SparkFun (9457) |
| Adafruit I2C 4-Digit, 7-Segment Display Module | Adafruit (878) |
| Adafruit Ultimate GPS Breakout | Adafruit (746) |

# Electronic Components

A kit of the parts listed next, among other things, and designed specifically for this book is available from www.monkmakes.com.

| Part | Suppliers (Product Code) |
|---|---|
| 470 Ω Resistor | Mouser (293-470-RC), CPC (RE03840) |
| 1 kΩ Resistor | Mouser (293-1k-RC), CPC (RE03803) |
| 10 kΩ Trimpot | Adafruit (356), SparkFun (COM-09806), Mouser (652-3362F-1-103LF), CPC (RE06540) |
| Photoresistor | Adafruit (161), SparkFun (SEN-09088), CPC (RE05271) |
| 2N3904 Transistor | SparkFun (COM-00521), Adafruit (756), CPC (SC10960) |
| TIP120 Transistor | Adafruit (976), CPC (SC10999) |
| 1N4001 Diode | SparkFun (COM-08589), Adafruit (755), CPC (SC07332) |
| Red LED | SparkFun (COM-09590), Adafruit (299), CPC (SC08044) |
| RGB Common Cathode LED | SparkFun (COM-11120) |
| TMP36 Temperature Sensor | SparkFun (SEN-10988), Adafruit (165), CPC (SC10437) |
| Tactile Push Switch | SparkFun (COM-00097), Adafruit (504), CPC (SW04726) |

# Tools and Prototyping

| Part | Suppliers (Product Code) |
|---|---|
| Male-to-Male Header Leads | M-M jumper wires: SparkFun (PRT-08431), Adafruit (759) |
| Half-Breadboard | SparkFun (PRT-09567), Adafruit (64) |
| Multimeter | Adafruit (71), SparkFun (TOL-09141) |
| Soldering Kit | SparkFun (KIT-11210), Adafruit (136) |

# Miscellaneous

| Part | Suppliers |
|---|---|
| Mini Servo | SparkFun (ROB-09065), Adafruit (1449) |
| 6V Small DC Motor | Adafruit (711) |
| 5V Power Supply | Adafruit (276) |
| 12V 2A DC Power Supply | Amazon.com |
| Adafruit Wi-Fi Dongle | Adafruit (814) |
| 5V 500mA USB Backup Battery | eBay, Amazon |
| Barrel-Jack-to-Screw Terminal Adapter | Adafruit (368) |
| 12V 5W Automotive Lamp | Supermarket, motor accessories retailer |
| Solid-Core Insulated Wire | SparkFun (PRT-08024) |
| Magician Robot Chassis | SparkFun (ROB-10825) |

# B

# JavaScript Quick Reference

**This appendix** is not a complete JavaScript reference; rather, it provides a quick reference for the main features of the language in tabular form.

## Core Language

|  | Examples | Notes |
|---|---|---|
| Declaring a variable | var x;<br>var x = 10;<br>var y = "Hello";<br>var z = true | Caution: if "var" is omitted, the variable is created with global scope even when defined in a function. |
| Looping | for (var x = 0; x < 10; x++) {<br>  // do something<br>} | This example counts from 0 to 9 using x as the counter variable. |
| Ifs | if (x > 10) {<br>  // do things<br>}<br>else if (x > 5) {<br>  // do other things<br>}<br>else {<br>  // do something else entirely<br>} | The "else if" and "else" clauses are both optional. |
| Comments | // line comment<br>/*<br>Block comment<br>*/ |  |
| Defining a function | function doSomething(param) {<br>}<br>var doSomething = function(param) {<br>} | These two examples are both valid. |

| | Examples | Notes |
|---|---|---|
| Calling a function | doSomething(x);<br>y = doSomething(x); | The keyword "return" is used to specify a return value. |
| Timeouts | setTimeout(myFunction, 1000);<br>setTimeout(function() {<br>  console.log("Boo!");<br>}, 5000); | The second parameter is the duration in milliseconds after which the one-off timeout should occur. |
| Repeat timers | setInterval(myFunction, 1000);<br>setInterval(function() {<br>  console.log("Boo!");<br>}, 5000); | The first example runs "myFunction" every five seconds. The second example uses an anonymous function. |

# Strings

| | Examples | Notes |
|---|---|---|
| Constants | var x = "abc";<br>var x = 'abc'; | Single or double quotes can be used to define string constants. |
| Concatenation | var x = "abc";<br>var y = "def";<br>var z = x + y;<br>'abcdef' | Note that numbers can also be concatenated to a string, but the first item must be a string. |
| Character access | var x = "abc";<br>x[0];<br>'a' | The item returned is a single character string. |
| Length | var x = "hello world"<br>var y = x.length; | This sets y to the length of x (11). |
| Replacing | var x = "hello world"<br>'hello world'<br>x.replace("world", "simon")<br>'hello simon' | Replaces all matches of the first parameter with the second. |
| Converting to numbers | var s = "-123.45";<br>var x = parseFloat(s); | You can also use "parseInt" if you just want the whole number. |
| Changing case | var s = "hello world"<br>s = s.toUpperCase();<br>s = s.toLowerCase(); | |
| Tokenizing | var s = "field1, field2, field3";<br>var tokens = s.split(','); | Create an array of elements, using the comma (,) as a field separator. |

# Arrays

|  | Examples | Notes |
|---|---|---|
| Creating | var x = ['abc', 123, true];<br>var x = []; | The first example initializes the array with three elements. |
| Getting elements | var x = ['abc', 123, true];<br>var y = x[0]; |  |
| Setting elements | var x = ['abc', 123, true];<br>x[1] = 234; | Setting an element outside the bounds of the array increases the size of the array as needed. |
| Length | var x = ['abc', 123, true];<br>y = x.length; | In the second example, y will have the value 3. |
| Appending | var x = ['abc', 123, true];<br>x.push('end'); | The string 'end' will be appended to the array. |
| Splicing | var x = ['abc', 123, true];<br>x.splice(1, 0, 'end'); | Both are used to add and remove elements from a list. The parameters to "splice" are index position, number of elements to remove, and an optional number of elements to add at that position. |
| Sorting | var x = [4, 8, 1, 6, 5];<br>x.sort(); | Sort ascending. |

# Objects

|  | Examples | Notes |
|---|---|---|
| Creating | var x = {};<br>var x = { fred: 123, jane: 456 }; |  |
| Adding elements | x['mike'] = 789;<br>x.mike = 789; | Both of these examples do the same thing.<br>As well as adding, they will also change an existing value. |
| Getting elements | var number = x.mike;<br>var number = x['mike']; | Both of these examples do the same thing. |
| Converting to a JSON string | var x = { fred: 123, jane: 456 };<br>var s = JSON.stringify(x); |  |
| Creating an object from JSON | var s = '{"fred":123,"jane":456}'<br>var x = JSON.parse(s); |  |

# Math

|  | Examples | Notes |
|---|---|---|
| Arithmetic operators | +, -, *, /, % | % is modulo remainder. |
| Math.abs | y = Math.abs(-23) | Returns the unsigned value. |
| Math.sin, also cos, tan, asin, acos, atan | Math.sin(Math.PI / 2); | Trigonometric function. Angles are in radians. |
| Math.pow | Math.pow(2, 10); | Raises the first parameter to the power of the second. |

For a full list of math functions, see http://www.w3schools.com/jsref/jsref_obj_math.asp.

# Dates

|  | Examples | Notes |
|---|---|---|
| Current date/time | x = new Date(); | Uses the BBB system time, which you may need to set using this: # ntpdate -b -s -u pool .ntp.org |
| Date | x.getFullYear(); // E.g. 2014<br>x.getMonth(); // month of year 0 to 11<br>x.getDate(); // day of the month 1 to 31<br>x.getDay(); // Day of weak 0 to 6 (Sun 0) |  |
| Time | x.getHours(); // 0 to 24<br>x.getMinutes(); // 0 to 59<br>x.getSeconds(); // 0 to 59 |  |

For more information on dates, see http://www.w3schools.com/jsref/jsref_obj_date.asp.

# C

# BeagleBone Black GPIO Pinout

You will probably want to keep this diagram on hand when planning an electronics project with the BeagleBone Black (see Figure C-1). The pin identifier is shown in the larger font (for example, P9_11) and subsidiary roles such as RX, TX, SCL, SDA are shown in a smaller font.

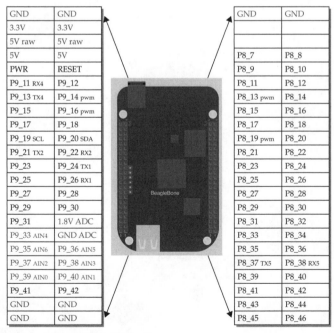

| | | | | |
|---|---|---|---|---|
| GND | GND | | GND | GND |
| 3.3V | 3.3V | | | |
| 5V raw | 5V raw | | | |
| 5V | 5V | | P8_7 | P8_8 |
| PWR | RESET | | P8_9 | P8_10 |
| P9_11 RX4 | P9_12 | | P8_11 | P8_12 |
| P9_13 TX4 | P9_14 pwm | | P8_13 pwm | P8_14 |
| P9_15 | P9_16 pwm | | P8_15 | P8_16 |
| P9_17 | P9_18 | | P8_17 | P8_18 |
| P9_19 SCL | P9_20 SDA | | P8_19 pwm | P8_20 |
| P9_21 TX2 | P9_22 RX2 | | P8_21 | P8_22 |
| P9_23 | P9_24 TX1 | | P8_23 | P8_24 |
| P9_25 | P9_26 RX1 | | P8_25 | P8_26 |
| P9_27 | P9_28 | | P8_27 | P8_28 |
| P9_29 | P9_30 | | P8_29 | P8_30 |
| P9_31 | 1.8V ADC | | P8_31 | P8_32 |
| P9_33 AIN4 | GND ADC | | P8_33 | P8_34 |
| P9_35 AIN6 | P9_36 AIN5 | | P8_35 | P8_36 |
| P9_37 AIN2 | P9_38 AIN3 | | P8_37 TX5 | P8_38 RX5 |
| P9_39 AIN0 | P9_40 AIN1 | | P8_39 | P8_40 |
| P9_41 | P9_42 | | P8_41 | P8_42 |
| GND | GND | | P8_43 | P8_44 |
| GND | GND | | P8_45 | P8_46 |

**Figure C-1**  *BeagleBone Black GPIO pinout*

# Index